Implementation Strategies for SAP R/3 in a Multinational Organization:

Lessons from a Real-World Case Study

Chetan S. Sankar
Auburn University, USA

Karl-Heinz Rau
Pforzheim University, Germany

CYBERTECH PUBLISHING
Hershey • London • Melbourne • Singapore

Acquisitions Editor: Michelle Potter
Development Editor: Kristin Roth
Senior Managing Editor: Amanda Appicello
Managing Editor: Jennifer Neidig
Copy Editor: Killian Piraro
Typesetter: Jessie Weik
Cover Design: Lisa Tosheff
Printed at: Yurchak Printing Inc.

Published in the United States of America by
CyberTech Publishing (an imprint of Idea Group Inc.)
701 E. Chocolate Avenue
Hershey PA 17033
Tel: 717-533-8845
Fax: 717-533-8661
E-mail: cust@idea-group.com
Web site: http://www.cybertech-pub.com

and in the United Kingdom by
CyberTech Publishing (an imprint of Idea Group Inc.)
3 Henrietta Street
Covent Garden
London WC2E 8LU
Tel: 44 20 7240 0856
Fax: 44 20 7379 0609
Web site: http://www.eurospanonline.com

Library of Congress Cataloging-in-Publication Data

Sankar, Chetan S.
Implementation strategies for SAP R/3 in a multinational organization : lessons from a real-world case study / Chetan S. Sankar and Karl Heinz Rau.
p. cm.
Summary: "This book presents an instructive insight into the complex process of ERP implementation in a global company"--Provided by publisher.
Includes bibliographical references and index.
ISBN 1-59140-776-1 (hardcover) -- ISBN 1-59140-777-X (softcover) -- ISBN 1-59140-778-8 (ebook)
1. Production management--Data processing. 2. Production planning --Data processing. 3. SAP R/3. I. Rau, Karl-Heinz. II. Title.
TS155.6.S246 2006
650.0285'53--dc22

2006007595

British Cataloguing in Publication Data
A Cataloguing in Publication record for this book is available from the British Library.

Implementation Strategies for SAP R/3 in a Multinational Organization: Lessons from a Real-World Case Study

Table of Contents

Foreword

Dear Reader:

Having taught and counseled thousands of students as a faculty member, graduate dean, and academic vice president before becoming involved with SAP, I know that the first question you will ask about the Robert Bosch case study presented in this book is, "Why is this important?" The second is, "Why do I need to know this?" Let me answer both of these questions by telling you what corporate leaders are thinking about and what keeps them awake at three in the morning.

In the spring of 2004, IBM conducted a massive global survey of CEOs. Over 450 CEOs from large, medium, and small enterprises across many industries responded. When asked what would have the greatest impact on their organization over the next three years, an overwhelming majority of the CEOs identified continuous changes in market forces to have the greatest impact. They talked about the intense and ever-increasing level of global competition, continuous changes in market dynamics, and rapid changes in customer needs and wants as the major market forces. Very few of them said they considered their organizations to be ready to respond to these changes, and most talked about responsiveness, agility, and flexibility as the new key competencies needed to survive and thrive in today's business environment.

These results are reinforced by the results of the 2004 Conference Board CEO Challenge survey. This survey asks CEOs to identify the top ten challenges that they face. "Speed, flexibility, adaptability to change" was identified as a challenge by 42% of the 539 CEOs responding to the survey, ranking it second only to "Sustained and stead top line growth."

The corporate drama in which you are about to be immersed as you read through this book illustrates how responsiveness and agility are important factors in high-cost, potentially high-risk technology decisions. However, there is another dimension. In his book *Adapt or Die*, Claus Heinrich (2003), a member of the SAP Executive Board, describes how adaptive business networks can help businesses meet the extraordinary agility challenges of the twenty-first century and what they will need to change in order to survive: "Falling margins, accelerating innovation and production cycles, as well as globalization are forcing companies to become more flexible in order to meet these challenges." He explains why this new business model is necessary for survival — and not merely an option.

A central, critical issue must also emerge from your examination of the Robert Bosch case presented in this book. Increasingly, *business process drives technology decisions*. Dr. Tom Davenport, professor and research director of the Institute for Process Management at the Babson Center for Executive Education, identified important insights from a study he coordinated in his previous role as director of the Accenture Institute for Strategic Change. Tom's research group surveyed "C" level officers of corporations that had recently implemented ERP systems. When asked why the company decided to implement an ERP system, the number one reason was to make better business decisions. Business decisions derive from and drive business processes. ERP implementations are intended to support and provide for corporate flexibility and profitability. In fact, a study of over thirty companies implementing ERP systems indicated that the market value of the stock of these companies increased just when they announced they were planning an ERP implementation.

The initial results of an ongoing global survey of over 600 companies has indicated that nearly 60% of companies surveyed believe that new business models (i.e., how the business is run) will be a greater source of competitive advantage than new products and services over the next 5 years, with nearly 82% of respondents believing that technology will be critical to the organization's ability to adapt the business model and implement strategy between now and 2010.

The real case study presented in this book describing the Bosch implementation of SAP starkly and undeniably weaves together the facts that business processes are integrated and that business process models drive technology decisions. To you as professionals, either presently or soon to be, in the twenty-first century corporate workforce, these are two critical lessons to learn.

Finally, in her keynote talk at the Gartner Symposium, Carly Fiorina, chairman and CEO of HP, citing Darwin's theory of evolution, said, "The company that thinks it's done is done. Darwin said it's not the most intelligent or strongest that survive, but those that adapt the most readily to change. It's not about fixing a company and stopping it, but it's about a company being able to adapt." As you examine the case presented in this book, seek out the critical lessons, remember them, and use them as guides. If you do this, then you will be prepared to adapt, to survive, and to contribute to your own all-too-real business cases of the future.

Sincerest best wishes,

Daniel C. Pantaleo, Vice President
SAP America, Inc., New Town Square, PA
February 2006

* * * *

It is my pleasure to write a foreword to this book by Drs. Karl-Heinz Rau and Chetan S. Sankar that documents in an excellent way how SAP R/3 was implemented in our company during the past ten years. This project has been exciting and challenging and we are pleased that this team has documented it thoroughly and well. I expect this book to be very useful to managers of companies who are in the process of implementing Enterprise Resource Planning Systems (ERP). In addition, it will be very helpful to students who are graduating with a major in information systems.

I have often been asked why Bosch has spent millions of Euros implementing a new system. Reading this book will hopefully provide you with the answer to this question and give you an in-depth understanding of the trials and tribulations involved in implementing such a system in a multinational company. I am proud to work for Bosch, an innovative and progressive company that was founded by Robert Bosch with great values and a strong mission statement more than 125 years ago. Richard von Weizsaecker, a former president of Germany, said on the occasion of the 125th anniversary of Robert Bosch's birth: "Robert Bosch is among the very best that Swabia has produced."

The success story reported in this book could not have been achieved without the exceptional engagement of the executive management and the experienced staff at QI in Europe and their counterparts in the regions. We hope that you will gain positively by reading about the change in management that has led to the successful ERP implementations in both our European and U.S. operations. We have approved the case studies that are discussed in Chapters VI, VII, and IX for use in classrooms and in training sessions. Please note, however, that the book is limited in that it does not provide all the detailed change processes that went through the company and only provides a series of snapshots of the situations at particular moments of time when the faculty members visited with us and gathered information. Hopefully our company's sharing of the implementation of SAP R/3 through this book will motivate you to become an ERP professional.

The concepts discussed in this book will become even more important in the future, as the usage of integrated IT-Systems for ERP are a prerequisite for making worldwide operations more efficient and cost effective. I ask you to read through this book, reflect on the lessons learned, and be prepared to improve the processes when your company starts implementing SAP R/3 or a competitor's ERP system. There are many benefits that would accrue to your company if your company is willing to work through the process and implement such a system.

Gerd Friedrich, Chief Information Officer
Robert Bosch GmbH
Stuttgart February 2004

Preface

The purpose of this book is to assemble many of the theories related to implementing an Enterprise Resource Planning (ERP) system, provide the history of implementation of SAP R/3 system in a multinational company during 1991-2004, and analyze the practice using the theories. We include a chapter that discusses the new products of SAP and shows how they might be used by the company in the future. We have also included another case study that discusses implementation of SAP R/3 in Sidler GmbH. This book provides the basics of ERP systems, so that a reader who is not familiar with the terminologies can understand them well enough to appreciate the importance of effective implementation of ERP systems.

This book can be used as a reference by a seasoned practitioner at any level of an organization or as an introduction to ERP by a person who is new to the concepts. It is particularly useful for project teams that are selecting or implementing ERP systems because it provides a detailed case study of implementation of a system at a company. The team can analyze the reported implementation, come up with alternate scenarios, learn from the exercise, and then proceed to work on effectively implementing the ERP system in their company. They can learn from the successful strategies and avoid the mistakes.

The book is particularly useful to students who are planning to graduate with an emphasis on business processes/ERP systems and practice as consultants or project managers in ERP implementation projects. It is especially helpful for students in graduate MBA and Executive MBA programs since it prepares them to be in charge of ERP implementation projects. This book provides them an insight into the implementation experiences of a large corporation that is expecting to spend about $1 billion on an ERP project. This book is unique since it emphasizes and illustrates business processes, IT architectures, and IT structures.

Focus of This Book

The success of a company depends on people, products, and processes. Typically in the beginning there is an innovative idea to solve an existing problem or to create a new demand with a new product. Examples in history of people and processes coming together to develop innovations are numerous and in this book we will take a look at some of them.

In one example of an innovative idea being applied in a different field, Herman Hollerith decided to use the punch cards developed by the French silk weaver Joseph-Marie Jacquard to represent the data gathered for the American census of 1890, and to read and collate this data using an automatic machine (Maxfield & Montrose, 1998). This became the basis for the world's most famous information technology company, IBM.

In another example, two German engineers, Gottlieb Daimler and Karl Benz independently invented the engine and the automobile in the 1880s and laid the foundations for the motorization of road transport. With the help of financial backers and partners, they both turned their development projects into private businesses, Benz & Cie. in October 1883, and the Daimler-Motoren-Gesellschaft (DMG) in November 1890. In 1885, Benz designed and built the world's first practical automobile to be powered by an internal-combustion engine and received the first patent (DRP No. 37435) for a gas-fueled car on January 29, 1886. In June 1926, the two oldest motor manufacturers merged to form Daimler-Benz AG (DaimlerChrysler, 2005). DaimlerChrysler now has a global workforce and a global shareholder base. With 384,723 employees, DaimlerChrysler achieved revenues of EUR 142.1 billion (US$192.3 billion) in 2004.

In 1886, Robert Bosch opened his "Workshop for precision and electric engineering." The next year he met Gottlieb Daimler, and Bosch developed a low-voltage magneto for the internal-combustion engine in stationary machines for him (Heuss, 1994). From that humble beginning, Robert Bosch GmbH has grown to be a 40 billion Euro company in 2004, with about 240,000 associates worldwide. By 2004, the Bosch Group had become the world's largest automotive supplier in terms of sales (Robert Bosch, 2005).

Finally, we want to mention SAP, a software company. In 1972, five former IBM employees launched a company called SAP (Systems Analysis and Program Development) in Mannheim, Germany, the same city where Karl Benz had started his company almost ninety years earlier. Their vision was to develop standard application software for real-time business processing. One year later, their first financial accounting product was ready for the market, forming the basis for the continuing development of other software components in what later came to be known as the "R/1 system," where "R" stands for real-time data processing. By the end of the decade, an intensive examination of SAP's IBM database and dialog control system led to the birth of SAP R/2. By 2005,

12 million workers around the world were using SAP solutions every day. There are now 91,500 installations worldwide, more than 1,500 partners, over 25 industry-specific business solutions, and more than 26,150 customers in 120 countries. SAP is the world's third-largest independent software vendor (SAP AG, 2005).

These examples show that people and products are essential for the success of companies. In particular, familiar products and the people who are the driving force of a company are visible to outsiders. However, less obvious factors actually determine a company's success or failure, such as the operating structure and processes they use. This is the focus of our book.

We will examine the internal and external processes used in a company and the information systems that enable and support these processes. We will not necessarily present new models and theories; our contribution to the academic and professional community is to present and analyze three case studies on the implementation of Enterprise Resource Planning (ERP) Systems in multinational organizations.

Robert Bosch Case Study

The major case study discussed in the book is the implementation of the SAP R/3 system by Robert Bosch GmbH, the world leader in the automotive parts supply industry. If you visit the homepage of the Bosch Group, you will find the term "diversity" mentioned several times. In the Bosch environment, this means diversity in products and technology (automotive technology, consumer goods and building technology, and industrial technology), and diversity in locations (more than 260 production locations in more than 50 countries). It also means diversity and complexity in the structure and processes used in the different plants and divisions of this well-established company. SAP's software system, R/3, basically structures business functions such as accounting, sales, purchasing, and so forth. It has changed its concept in the last few years to focus on business processes. However, it will take more time to master this transition, as is demonstrated in the book. The company is still developing new technologies to build process-oriented software solutions. Leading IT providers are still undergoing a transition process as they begin to provide new solutions that support a business process-oriented strategy.

The Robert Bosch GmbH case study in this book covers a period of more than 10 years. It starts in 1994, when the company's board of management established an IT division with a corporate-wide responsibility for IT. Traditionally, the IT functions in companies have a strong technology orientation. In the past few years, however, many companies have realized that IT solutions are closely connected with business processes. The design of business processes is essential for business success. The quality of business processes measured in cycle-time, reliability, flexibility, reactivity, and costs is highly dependent on the information systems that support them. The issues involved in bringing together information systems, business processes, and the people who will have to work with them is the main subject of this case. Theoretically, the solution should be relatively easy: Optimal processes are designed using the best information systems and then implemented. The problems arise when this is done in a real company with real people. To teach this to students in the classroom is very challenging. Students typically have no idea about the complexity of real companies. To describe this in an abstract way is either trivial or impossible, so we decided to use the case study approach. We did not try to simplify and structure reality, since it is neither simple nor well structured. These are the types of challenges our graduates will face once they leave the classroom and enter the workforce.

When we talked to the IT executives at Robert Bosch in 1999, they were in the middle of the transition process as they reevaluated the way they were doing business. We learned that SAP was being implemented in Robert Bosch worldwide, but that the top management of Bosch was not satisfied with the implementation of more than 50 SAP R/3 systems. The case study shows that it took some time to convince the top management, as well as the people in the plants and departments, what needed to be done. The challenges were first to get the support of the company's top management for the necessary changes in the business processes. On the operational level, it was necessary to convince the workforce that an optimal solution in a certain plant may not be the optimal solution from a corporate point of view. It was necessary to make clear that standardization and harmonization of business processes would provide benefits for the individual plants and business units, as well as for the corporation as a whole. We also had the chance to document Bosch's existing, and very innovative, concepts for providing optimal integration of IT systems in 1999, which turned out to be very similar to the new concepts that are now included in the NetWeaver-based integration of SAP. The first two case studies developed during this period showed the managerial and technical issues that had to be solved for effective implementation of the SAP R/3 system. Further discussions during 2004 led to the development of an extra chapter on how Robert Bosch implemented the SAP R/3 system, along with documentation of the change in management processes that were put into practice.

To round out the book and provide another perspective, we also developed another case study by working with the management of Sidler GmbH in Germany on the implementation of SAP R/3 in their company.

We have used all these case studies in graduate and undergraduate business classes at Pforzheim University in Germany and Auburn University in the United

States, where they were very well received by the students. Even those students who had taken a few courses on SAP R/3 told us that they understood for the first time the business implications of implementing SAP R/3 in a multinational corporation when they analyzed the Robert Bosch case study.

Organization of This Book

As we developed the case studies, we worked with Idea Group Publishers to produce the case studies in the form of a textbook. They provided us with feedback that convinced us that it was essential to include chapters that provide general information about multinational companies, ERP systems, change management topics, and the technical issues involved. We therefore added several introductory chapters to make it easier for students to understand and benefit from analyzing the case studies. The book is presented using twelve chapters, and we will summarize the contents of each chapter briefly below.

Chapter I introduces the book and explains the pedagogical approach we follow, particularly how the case study approach can be used to bring real-world issues into the classroom. We make especially clear how the learning process works. This should help the instructor using this book, as well as help the students learn how to use the case studies and derive the most value through analyzing them.

Chapter II defines and describes how multinational companies function. We explain the basic features of organizational structures and how large companies structure their divisions. In addition, we explain the strategies followed by multinational companies to manage their divisions and plants.

Chapter III provides a general overview about Enterprise Resource Planning systems, starting with historical approaches such as MRP and MRP II. A brief description of the functions or modules of an ERP system helps students to understand the strength and complexity of such a software solution. Due to the fact that an ERP system is not a simple software solution, such as a spreadsheet system used on a single computer by one user, we provide a short description of the technical architecture of an ERP system. An overview of the ERP market is also provided and the major companies that sell ERP systems are briefly described. Due to the fact that the Robert Bosch case uses the SAP R/3 system, we go on to describe SAP and their ERP system in detail. Using this chapter, a reader with only a limited background in ERP can gain a general idea about this subject. An experienced reader could read this chapter quickly or even skip it.

Chapter IV introduces the concept of change management. To implement an ERP system company wide means that it will be necessary to manage the change process. Basically, this is not a technical issue but a psychological challenge, and the reader is given a well-founded overview of the change management life cycle in this chapter. Learning about the life cycle is essential for future executives who will have to manage changes in organizations. Based on this understanding of an ERP system, we can then explain the issues involved in implementation, concentrating on structural, technical, and managerial issues and presenting a range of possible implementation strategies and approaches.

Chapter V describes the basic concepts of database management systems and defines terms such as "normalization" and "single instance," with examples. The chapter describes the issues that must be considered when implementing ERP systems and the phases that could be used to implement an ERP system. It concludes by describing enterprise modeling methodology and how a company could be structured.

Chapter VI is the first case study, covering the period 1992-1999. It also provides an overview of the Bosch company to give the reader a picture of the different business sectors of Bosch and how Bosch is structured. The reasons for implementing a corporation-wide IT division are also discussed. The rest of the chapter focuses on understanding the business, the structure, and the IT issues within Robert Bosch in the United States (RBUS). In this chapter, the reader will also meet the people who were the main players in implementing the SAP R/3 system at Robert Bosch.

Chapter VII is the second case study on Robert Bosch, and shows how a solution for the IT issues in Robert Bosch US could not be approved without finding a solution for the whole company, due to the fact that Bosch does its business on a global scale. We present the process followed by the CIO at RBUS as he worked on a proposal to solve the problems in Robert Bosch U.S., while at the same time the central IT division in Germany was working on concepts to standardize and harmonize the company's business processes, as well as defining a standardized systems solution. The top management of Robert Bosch US initially rejected the project proposal presented by the CIO at RBUS, since he proposed using a specialized solution in the US, which would not be suitable for the whole company. These two case studies provide students with an opportunity to propose future strategies for implementing SAP R/3 throughout the corporation.

Chapter VIII provides an analysis of the case studies performed by the authors on the case studies presented in Chapters 6 and 7. Analytical methods are applied to produce answers to the questions raised in the case studies. This chapter can be used in different ways: The first approach would be for the instructor to use these analysis to prepare student assignments, and the second approach would be for the reader to compare his or her own analysis to those provided by the authors.

Chapter IX continues the case study provided in Chapters VI and VII, covering the period from the beginning of 2000 to the summer of 2004, including the company's projections out to 2008. It shows the reader what Robert Bosch did in this period to implement an ERP solution for a company of this size. This SAP R/3 project is one of the largest undertaken anywhere in the world, and is expected to cost more than one billion dollars. Therefore, the reader has the unprecedented benefit of learning from an exceptional case study. This chapter presents changes that are not limited to technical aspects, but affect the way Robert Bosch is managing its business globally. But it also shows that in a company of this size, a step-by-step implementation approach is the only way to execute such a major change successfully. Nevertheless, this approach does not lead to an optimal solution in an ideal sense.

Chapter X provides the reader with some idea of the future of ERP solutions, which are expected to provide businesses with a high degree of integration and flexibility. The authors of this chapter are working at the forefront of the future software solution architectures, and collaborate closely with SAP. The material describes the basics of the Service Oriented Architecture (SOA), which was mentioned as the future direction in Chapter 3. Based on this general understanding, the Enterprise Service Architecture (ESA) is presented. The description is not limited to concepts, but shows also how this ESA concept is supported by NetWeaver, which is the product portfolio of SAP that enables flexible integrated solutions. This chapter shows that many of the problems described in the case studies will be solved more elegantly in the future, and it shows also how these technologies will provide countless new possibilities to support business processes in more optimal ways. Many of the described concepts and products are still under development, but they are already in place and used by some companies.

Chapter XI delivers an additional example of ERP implementations in another company. The Sidler company is also active in the automotive supplier industry. It is a fairly small company, but also works in different parts of the world and supplies car manufacturers all over the world. They had to implement a new ERP infrastructure because their existing IT solutions were outdated and were not able to comply with theY2K issue or to cope with the introduction of the new European currency, the Euro. They had a low level of expertise in managing IT projects of this size, so they had to rely on external consultants. This led to many problems, but finally they implemented the system successfully. This was basically possible because the size of the company was limited and they were able to improvise.

Chapter XII concludes the book and summarizes the benefits the reader will have gained. It reviews the inventory of skills that a student should have improved by analyzing the case studies and lists lessons in implementing ERP systems that could be applicable to the student's future career. This chapter analyzes the change management processes adopted by Robert Bosch using the change management life cycle theory. In particular, it provides students with an in-depth understanding of ERP implementation by analyzing the implementation of SAP R/3 systems at Robert Bosch RB GmbH during 1991-2004. Overall, the book identifies key management issues in designing and implementing ERP systems and illustrates them with real-world examples.

Strength of This Book

The strength of this book lies in the combination of the theories that are provided in Chapters I through V and the case studies that illustrate the theories in Chapters VI through X. Other chapters analyze the case studies and connect the theories with the practical examples. This process gives students an opportunity to understand how to apply the theories they learn in class to a practical problem. The students will inevitably face unsolved and difficult problems in the future, and we expect the approach used in this book will provide them with the valuable analytical skills they will need to address these problems.

Reference List

DaimlerChrysler. (2005). The Brand Mercedes-Benz. Retrieved August 15, 2005, from https://www.cms.daimlerchrysler.com/emb_classic/0,,0-195-78709-1-84456-1-0-0-0-0-0-434-78641-0-0-0-0-0-0-0,00.html

Heuss, T., Gillespie, S., Kapczynski, J. (1994). *Robert Bosch: His life and achievements*. New York: Henry Holt & Co.

Maxfield & Montrose Interactive, Inc. (1998). A history of computers: 1890 AD Herman Hollerith's tabulating machines. Retrieved August 15, 2005, from http://www.maxmon.com/punch1.htm

Robert Bosch GmbH. (2005). Structure and business sectors. Retrieved August 15, 2005, from http://www.bosch.com/content/language2/html/2143.htm

SAP AG. (2005). SAP history: From start-up software vendor to global market leader. Retrieved August 15, 2005, from http://www.sap.com/company/history.epx

Acknowledgments

The work on this book goes back to fall 1999, when we first contacted Dr. Eggensperger, the CIO of Robert Bosch GmbH in Stuttgart, Germany, and the idea to write a case study about the ERP implementation process in the Bosch corporation was born. The authors decided to focus the case study on two aspects: the corporate view and the view of the Bosch subsidiary in the United States of America. In this first phase, we interviewed different IT executives in the U.S. as well as in Germany. We would like to express our gratitude to Mrs. Margit Bauer, CIO Robert Bosch GmbH, and Mr. Don Chauncey, vice president of Information Systems Robert Bosch US, for providing us with invaluable information. Based on this field research, we were able to prepare two case studies. For contributing to the process of improving the cases studies, we also thank the reviewers of the Ninth Annual International Casewriters' Workshop held during the meetings of the World Association for Case Method Research & Application (WACRA) held in Budapest, Hungary, July 2000. In the semesters following, we used the case studies in different undergraduate and graduate classes, and we appreciate the advice and feedback given by our students at Auburn University and Pforzheim University.

In 2004, we decided to publish the case studies, supplemented by additional material, as a book. In this context we contacted the Bosch company again. In the summer of 2004, we conducted a series of interviews with Mike Bieganski, vice president of Information Technology for Robert Bosch North America in Chicago, and Gerd Friedrich, CIO of Robert Bosch GmbH in Stuttgart. We appreciate the information that they shared with us, which enabled us to conduct a longitudinal study covering a timeline of about 10 years. We are also grateful to Mr. Friedrich for writing a foreword for this book expressing the importance of publishing books that combine theory and practice.

In February 2001 and March 2005, we had the chance to discuss our results with MIS faculty at the SAP Research and Application Congress 2001 in San Diego, California, and the SAP Curriculum Congress 2005 in Atlanta, Georgia. We again received valuable feedback that we were able to use to improve our cases. In this context, we are much obliged to Dan Pantaleo, vice president of SAP, who also wrote a foreword for this book expressing how the lessons learnt from the case studies presented are so helpful for professionals in the business arena.

In preparing a book, the professionalism of the experienced staff at the publisher is indispensable. We acknowledge the help of all involved in the collation and review process for this book. Kristin Roth, development editor, Jan Travers, senior managing editor, and Dorsey Howard, marketing assistant at Idea Group Publishing helped us tremendously in revising and finalizing the manuscript. Last but not least, our deep gratitude is due to the reviewers who gave us valuable advice to strengthen the value of the book.

We also thank our graduate students, Keegan B. Corcoran, Manuel Offermanns, Ramin Chandhok, and Nazmi Doganc who helped develop some of the materials used in the chapters. Van Norris helped by providing valuable comments on an earlier draft. Jan Szechi edited the material to make it more presentable. We are grateful to our colleague William R. Boulton for providing permission to use the Sidler GmbH case study. We also acknowledge the managers from this company, who provided permission to publish this case study. We are also thankful to our graduate students who provided feedback on earlier versions of the case study.

In closing, we wish to thank the authors of Chapter 10, Valentin Nicolescu, Holger Wittges, and Helmut Krcmar, from the Technische Universitaet Muenchen, for writing this chapter. Finally, we want to thank our respective spouses, Lakshmi and Jutta, for their love and support throughout this project.

Dr. Chetan S. Sankar

Auburn University, USA

sankacs@auburn.edu

Dr. Karl-Heinz Rau

Pforzheim University, Germany

karl-heinz.rau@hs-pforzheim.de

Chapter I

Introduction

Learning Outcomes

- Identify why Enterprise Resource Systems (ERP) have emerged as a critical success factor for companies
- Understand that the market for ERP systems is growing rapidly
- Know that skills to effectively implement ERP systems are valued by industry
- Understand the rationale supporting the selection of case methodology for this book
- Know how to use individual preparation, small group discussion, and large group discussion to analyze the case study
- Understand the organization of the book

Key Challenges Facing IT Professionals

Information Technology (IT) is one of the key factors driving progress in the 21st century — it is transforming the way we live, learn, work, and play. Advances in computing and communications technology have created a new infrastructure for business, scientific research, and social interaction. This expanding infrastructure provides us with new tools for communicating worldwide and for gaining knowledge and insight from this rapid flow of information. Information technology helps us understand how we affect the natural environment and how best to protect it. It provides a vehicle for economic growth. Information technology is making the workplace more rewarding, improving the quality of health care, and making government more responsive and accessible to the needs of our citizens. The world risks losing the scientific, economic, and human resource advantages it now enjoys without an IT workforce that is large enough to meet both the public and private sectors' growing demand, and is adept at using and producing information technologies.

Although information systems create many exciting opportunities for businesses, they are also a source of new problems, issues, and challenges for managers. Large software and hardware systems continue to fail despite rapid advances in information technology. For more than a decade, businesses and industries have complained that the skills of information systems (IS) personnel entering the workforce are not sufficient to meet the challenges of a high-performance workplace[1].

It is critical that information technology (IT) professionals and students are able to understand how IT can be used effectively in global organizations. The challenges companies face as they use information systems to make their operations much more responsive and effective are (Laudon & Laudon, 2004):

1. *The Strategic Business Challenge*: Despite their substantial investments in information technology, many organizations are not realizing significant business value from their systems, nor are they becoming digitally enabled. To fully benefit from information technology, realize genuine productivity gains, and take advantage of a company's full capabilities, many organizations may need to be redesigned, requiring changes in business models and organizational structures.
2. *The Globalization Challenge*: The rapid growth in international trade, partnerships among countries such as NAFTA, the European Union and ASEAN, and the emergence of a global economy highlights the need for information systems that can support production, marketing, and sales of goods in many different countries. There is an increasing need for transnational business processes and supporting information systems that can produce profit and loss and balance sheets for a multinational corporation on a daily basis, provide effective global logistics operations, and continue to perform effectively when companies within the corporation merge with new companies or are divested.

3. *The Information Architecture and Infrastructure Challenge*: Many companies are saddled with expensive and unwieldy information technology platforms that cannot adapt to innovation and change. Meeting new business and technology challenges may require redesigning the organization and building new information architectures and corresponding information technology infrastructures. Information architecture is the particular form that information technology takes in an organization to achieve selected goals or functions. It is essentially a design that shows how the firm's key business application systems are integrated together to meet the business needs. The firm's IT infrastructure provides the technology platform for this architecture. The computer hardware, software, data and storage technology, networks, and human resources required to operate the equipment constitute the IT infrastructure and should be available to all of its applications. However, most companies are crippled by fragmented and incompatible computer hardware, software, telecommunications networks, and information systems that prevent information from flowing freely between different parts of the organization.

Enterprise Resource Planning Systems to the Rescue

Enterprise Resource Planning (ERP) systems have emerged as a way to address the challenges posed above and are used to coordinate activities, decisions, and knowledge across many different functions, levels, and business units in a firm. Just as the Internet has revolutionized the way in which we gather and transfer information, ERP has the potential to revolutionize the fundamental ways in which information is stored and retrieved in an organization. Examples of information technologies that have effectively reshaped the business world include: TCP/IP and the World Wide Web, which have revolutionized how we make data accessible; Netscape and Internet Explorer, which have revolutionized the way in which we gather that data; and ERP systems, which are designed to standardize the manner in which transactions in a business are recorded and retrieved.

This book discusses the fundamentals of ERP systems and provides a detailed example of how Robert Bosch GmbH has used these systems to address the challenges. We expect that by reading through this book, you will realize that ERP systems are poised for explosive growth and use all around the world. Effective implementation of these systems can provide enormous benefits to companies and government. At the same time, it requires that you acquire a new set of skills to succeed in the IT profession, learn about change management principles, and understand how to design IT systems that facilitate effective change in organizations.

Market for ERP

The global revenues from the software used to automate financial, human resources, procurement, and other business operations (the ERP market) is expected to reach $26.7 billion during 2004, up from $25 billion in 2003 according to International Data Group. In addition, the top 10 ERP vendors collectively increased their market share during 2003 to 46% from 44% in 2002 and 42% in 2001. Of these, the top five vendors were SAP, PeopleSoft, Oracle, Microsoft, and Sage (TechWeb.com, 2004). The global enterprise resource planning (i.e., ERP) market is on the rise and this is likely to continue at least through 2009, with SAP and Oracle leading the revenue stream, according to a report by AMR Research. "Market Analytix Report: Enterprise Resource Planning, 2004-2009" shows that overall global growth was 14% in 2004, with 16% in North American, 12% in Europe, and 19% in Asia Pacific. However, when discrepancies in currency rates are taken into account, organic growth was closer to 8% or 9%, according to David O'Brien, vice president of quantitative research. Additionally, the report found, the European market is expected to grow from $8.8 billion in 2004 to more than $12 billion in 2009 (DeFelice, 2005).

Microsoft and SAP are cooperating in a 10-year agreement to integrate Microsoft's .Net with SAP's NetWeaver Platform. The stated goals are to make it easier for Microsoft Office users to connect to SAP applications and to make Visual Studio the development tool of choice for SAP connections (Fox, 2004, p. 20). The US market is the fastest-growing market within SAP (McDermott, 2004, p. 29). The number of job advertisements seeking to recruit staff with SAP skills rose by 53% during 2002, according to the SSP/Computer Weekly Survey (Langley, 2003, p. 60), and is ranked 17th in the CW/SSP list of top IT skills. Employers are expected to face shortages of IT professionals with enterprise resource planning skills during 2005-2006, and the demand for IT professionals with SAP, Oracle Finance and Siebel is growing rapidly, driven by a wave of technology upgrades by leading firms, public sector projects and financial compliance regulation, according to Parity Resourcing Solutions (Goodwin, 2005).

Not only are companies looking at ERP as a timesaving device, they are using it to save huge sums of money. Such radical improvements can only be achieved if a great deal of additional restructuring/reorganizing is done in a number of areas. This makes learning the fundamentals of ERP and learning about ERP implementations very important to you and to your future employers. So far, we have only mentioned the difficulties in implementation and how costly it can be. Even after a successful implementation, a company might face risks. These risks occur largely because of the complexity of ERP systems. We will also show that it is difficult to make things work seamlessly in a real organization.

Figure 1. Enterprise resource planning framework

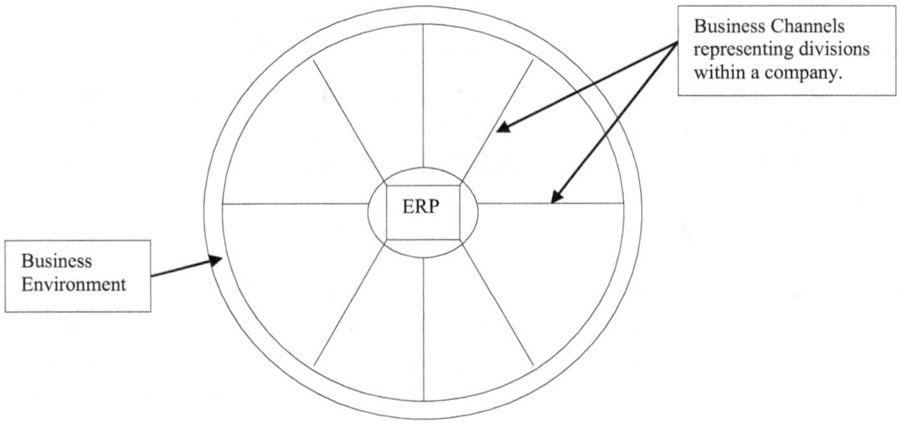

What is Enterprise Resource Planning?

Enterprise Resource Planning is a collection of software modules that integrate all of the divisions of a company via a centralized database (Larocca, 1999). This sounds simple, but it isn't. However, we can dissect the system into its components to see what it is and how it works. ERP systems are like a bicycle wheel (Figure 1). Think of the database as the center of the wheel, and give each division of the corporation a spoke that connects the center with the outside of the rim. Think of these outside connections at the rim as each division's point of contact with the business world. Due to the layout of ERP software, these divisions are usually Finance, Production, Marketing, Human Resources/Payroll, Accounting, and so on.

Prior to an ERP implementation, most divisions operated with relative autonomy and with little real-time interaction. This meant that the sales team may not know exactly what the production levels are, or what is actually in inventory. We say "actually" because even if the sales force has an idea of the inventory levels, if an order is processed, there is an immediate drop in actual inventory levels, although a time lag exists in the recalibration of the known inventory levels. For example, if I were to sell the last unit of product A and you were not aware of this, you would assume that unit was still in the inventory and you might also sell that unit, even though that unit is no longer available for sale. This predicament occurs in sales all the time, and the results are not just an embarrassing phone call for the salesperson, as the issue affects customer satisfaction/loyalty as well. Here is where ERP comes in. If I sell the last unit of Product A, it is recorded directly in the central database, where it is then taken off the current inventory level. Now, you receive an order for the last unit of product A. With the click of a mouse, you can see that unit is no longer in inventory and is therefore undeliverable, and suggest an alternative to the customer.

While this is a highly simplified example of how ERP works, it is possible to see the general concept behind what ERP does. It begins to get complicated once you start envisioning the inputs from all the divisions of a company and what the implications are. Here is a more elaborate example:

> *Let's assume that a South Africa-based salesperson for a U.S. multinational computer firm prepares a quote for a customer using an ERP system. The quote specifies a legally binding product configuration, price, delivery date, shipping method, and so forth, all determined in real time from the system. When the customer accepts the quote (via the Internet, let's say), a sales order is recorded. The system schedules the shipping (including shipping points and truck routes), then works backward from the shipping date to reserve material availability, order needed parts from suppliers, and schedule the computer assembly in manufacturing. The customer's credit limit is checked. The sales and production forecasts are updated. Material Requirements Planning (materials needed to produce the product and the sequence in which they need to be available) and bill-of-materials lists are created. The salesperson's payroll account is credited with the correct commission, and his or her travel account credited with the expenses of the sales call. Actual product cost and profitability are calculated. The division's and firm's balance sheets, accounts payable and receivable ledgers, cost center accounts, cash levels, and any other relevant finances are automatically recalculated in an instant. Virtually every information transaction resulting from the sale of the computer is taken care of except for the impact on the company's stock price.* (Davenport, 2000)

As you can see, the complexities now start to arise. But the principle is the same for every division, not just the sales division. Let's say that the Production Department is going to increase its production of a new product. This increase in the production schedule would then appear on the screen in the marketing department. The staff in marketing would know that they need to focus on moving this product because of the increase in volume. Finance would expect higher material costs as more material is ordered to produce at the new level, and the sales projections and accounting *pro formas* are all altered to indicate the new expectations.

But what happens when mistakes are made? Let us now suppose that someone in Production shows more units have been produced on a particular day than is actually the case, possibly because they are exaggerating the order to meet an end-of-month quota or they made a typo when keying in the number of units of inventory for Product A. Assume that this goes unnoticed by the managers concerned. The entry was for an extra 100 units of Product A, and each unit of this product has an inventory carrying cost of 10 dollars. Instantly the central database accepts this new inventory schedule and adjusts the number of units available for sale. Every salesperson's computer indicates that there are 100 units of product in inventory more than there actually are. Finance is also affected, as there is a debit of $1,000 to the cash flow for the inventory carrying costs of the product. This shows the ripple effects that a single error in one department can make. Sales may no longer match delivery schedules. Customers might become unhappy and losses may result. Because every department shares the same information, and bases all of their decisions on the data contained in the database, it is imperative that data entries are made with precision. Some of the major problems with implementing an ERP system lie here, as it is difficult to mandate the necessary level of precision because of the lack of checks and balances on employee data entry. A careless or malicious employee can conceivably cost the firm millions of dollars with just a few keystrokes. However, ERP systems are now sophisticated enough that they can issue a warning pointing out that it is not possible to have produced so many units based on the materials taken out of stock and committed to production. This impossible yield would be flagged and errors such as these should thus be caught and corrected quickly.

Benefits of an ERP System

It is clear that ERP systems can be very beneficial for companies and can increase their competitive advantage. As we have explained, an ERP system can create important business benefits. By integrating the whole business process, it optimizes it and increases efficiency. Of course, without cost reductions and good quality control, an ERP system cannot achieve its full promise. On the managerial level, by supplying the decision makers with appropriate and real-time data, it supports the decision making process in the company. In other words, it supports the decision makers at the strategic level. Probably the most appropriate benefit an ERP system brings is its integration potential. By integrating the business processes, it supports the organizational structure and helps to create a more disciplined organizational structure, cutting cycle times, and cost of business processes. By integrating geographically dispersed units, it helps to create more uniform organization structures. Also, from a technical point of view, implementing a single, unified and comprehensive information system technology means that the data is standardized and accessible for everyone in the company (Laudon & Laudon, 2004).

The question currently faced by managers and executives at many leading corporations is how to implement such systems successfully. The answer lies in the success of companies who have already gone through such an implementation and succeeded. For example: "IBM's Storage System (disk drive) division achieved a reduction in the time to enter pricing information from five days to five minutes, replacement part shipping went from twenty-two days to three, and credit checks that previously took twenty minutes are now accomplished in three seconds" (Davenport, 2000, pp. 7-8). If this increase in the speed of doing business fails to impress you, then take a look at Microsoft: "The company's ERP system has allowed it to receive \$14 million per year in early payment discounts from vendors" (Davenport, 2000, p. 8). Cisco has also realized substantial financial savings: "Cisco's system cost over \$15 million and the company spent another \$100 million connecting it to the Internet. Today, however, Cisco believes that the combination of its ERP system and its Internet applications yields more than \$500 million in annual operating cost savings" (Davenport, 2000, p. 8).

Challenges in Implementing ERP Systems

Although ERP systems can improve organization coordination, efficiency, and decision making, they have proven very difficult and costly to build. Many barriers must be overcome before the benefits of ERP systems can be fully realized (Robey, Ross, & Boudreau, 2002). A typical ERP system installation costs \$15 million and it may run to over \$100 million for very large companies who implement full-function systems across many divisions (Hitt, Wu, & Zhou, 2002). ERP software implementation is deeply intertwined with corporate business processes and it might take a large company three to five years to fully implement all the organizational and technology changes required. Because ERP systems are integrated, it is difficult to make a change in only one part of the business without affecting other parts as well. There is also the danger that the new ERP systems might eventually prove as brittle and hard to change as the old systems they replaced, binding firms to outdated business processes and systems.

Mergers and acquisitions (M&As) have proliferated in recent years because they are major growth engines for businesses. However, research has found that more than 70% of all M&As result in a decline in shareholder value, and often lead to divestiture at a later time (Frank & Sidel, 2002). Often, the failure can be traced to difficulties in integrating the information systems of different companies. Without a successful systems integration, the benefits anticipated from the merger cannot be realized or, worse, the merged entity can no longer execute its own business processes and might lose customers.

A high percentage of ERP projects fail to be fully implemented or to meet the goals of the users even after three years of work (Laudon & Laudon, 2004). Many ERP projects have been undermined by poor implementation and change management practices that fail to address employees' concerns about change. Study of failures of ERP systems in China reveal that ERP vendors might need to localize strategies. Several hurdles such as incompatible language, financial reporting changes, highly priced systems, and rapid restructuring of companies were identified as leading to failed implementations in China (Liang, Xue, Boulton, & Byrd, 2004).

Other researchers have pointed out that strategies such as sustained management support, effective organizational change management, and good project scope management and tactics such as dedicated staff and consultants, strong communication, and formalized project plans were some of the critical success factors in implementing ERP systems (Esteves-Souza & Pastor-Coalldo, 2000).

The above challenges make it important for you to learn how to implement ERP systems effectively in organizations.

Selection of the Case Study Method as a Way to Learn How to Implement ERP Systems

The methods that you can use to learn about effective ERP implementations are to: (a) read textbooks on ERP implementation, (b) obtain on-the-job training, and/or (c) use case studies. We will discuss these three methods and show why we chose to use the case study method to teach you about effective ERP implementations.

Read textbooks: There are many excellent books that describe how to implement SAP R/3 and other ERP systems (Bancroft, Seip, & Sprengei, 1997; Kale, 2000; Welti, 1999). They provide elaborate check lists and descriptions of the procedures that need to be followed in order to achieve an effective implementation. These check lists and procedures have been derived from the textbooks' authors' experiences and are invaluable. However, for a reader such as yourself, it is often difficult to figure out the relative priorities between the different topics covered in such books. These books might be very helpful in enabling those who are very familiar with ERP implementations to fine-tune their performance, but are of limited use in educating a person who has little or no previous experience in dealing with ERP systems.

A second way to learn about implementing ERP systems is to either intern at or obtain a full-time position with a company. However, there are only a few companies that have implemented large-scale ERP systems, and few of these will be willing to hire you unless you already have the prerequisite knowledge about ERP implementation. Even though this might be the best method, for most of you it is a rather difficult proposition to fulfill.

The third method is to learn about effective implementation of ERP systems by reading, analyzing, and critiquing an ERP implementation performed by a large multinational company over a period of many years. You can start to familiarize yourself with the implementation in this company by reading the materials by yourself, work in small teams to analyze the problem, and then work in large groups. This will enable you to analyze a real situation, commonly involving a decision, that was actually faced by individuals employed in the organization. The case study allows you to step figuratively into the shoes of the decision makers. The case study in this textbook is based on a billion dollar ERP implementation at Robert Bosch, GmbH, a large multinational organization. This case is field-based and we visited the organization several times to collect the data that comprises the case. The appropriate individuals at Robert Bosch have all signed an official release document. It is this release that truly distinguishes the case in this textbook from other kind of educational materials. The release serves four purposes: First, it guarantees that we in fact went into the field and did the work required for a complete and accurate study of the Bosch ERP implementation story, as opposed to simply writing about it from the comfort and security of our offices. Second, it authenticates the experience; a signed release says that the situation is accurately and fairly portrayed. Third, the release grants permission to use the case for educational purposes. Fourth, the release helps maintain positive relations between us and the IT managers at Robert Bosch (Mauffette-Leenders, Erskine, & Leenders, 1997a).

Definition of a Case Study and Its Purpose

Cases are descriptions of situations that frequently focus on a problem or a decision facing people in pursuit of their occupation or interests. They usually involve the consideration of an actual example, requiring the synthesis of a large amount of different kinds of information, and the making of recommendations or decisions (Kletz, 1970). The case study takes shape as the empirical details that constitute the decision or problem under study are considered in the light of the evidence gathered from the workplace (Naumes & Naumes, 1999). This gives depth and dimension to the scenario presented in the case study (Hamel, Dufour, & Fortin, 1993; Krone, Juedes, & Sitharam, 2002).

History of the Case Study Method

Case study materials can be used for a variety of purposes, such as conducting research, completing program evaluations, and as a teaching resource. The case study method first emerged as a method of teaching in the Harvard Law School in 1870, but was later formalized as an effective teaching method by the Harvard Business School. The written case study method is a well-established teaching method in many disciplines and fields of study. The origins of using the written case study as a teaching method in business schools can be traced to 1908, when the Harvard Business School (HBS) published the following statement in its catalog:

In the course on Commercial Law, the case-system will be used. In the other courses, an analogous method, emphasizing classroom discussion in connection with lectures and frequent reports on assigned topics, what may be called the "problem method" will be introduced as far as practicable. (Copeland, 1954)

Although the use of cases for teaching commercial law appeared logical, in 1911 the HBS extended the use of this method to the teaching of Business Policy. The unique feature of the course was the participation of fifteen businessmen in the instruction. Each businessman presented a problem to the class. At the first meeting with the class, he explained the problem to the class and answered questions. At the next meeting of the class two days later, each student handed in a written report embodying his analysis of the problem and his recommended solution. The reports were discussed by the businessman with the class in their third meeting.

The HBS soon had to resort to the written case method for teaching business policy for logistical reasons and due to the fact that the businessmen varied in their teaching abilities (Spizizen & Hart, 1985). Today, the written case study is a well-established teaching method in business schools across the U.S. for teaching not only graduate level courses, but also senior-level undergraduate courses. The University of Chicago used the case study method extensively in research on sociology during the early 1900s, but this method was later abandoned in favor of statistical methods developed at Columbia University.

A case typically is a record of a business or technical issue which has actually been faced by engineers and managers, together with the surrounding facts, opinions, and prejudices upon which engineers' and managers' decisions have to depend. These real and particularized cases are then presented to students for considered analyses, open discussion, and final discussion as to the type of action that should be taken (Bonwell & Eison, 1991; Daniels, Faulkner, & Newman, 2002; Garcia & Moreno, 2004; Lammers & Murray, 2002; Lamnias, 2002; Love, 2000; North American Case Research Association, 1997; Richardson, 2003; Seller, Turnwald, & Bull, 1994; Thorpe, 2000; Ul-Haq, Stiles, & Pond, 2003).

MIS Case Studies

We have modified the ideas used in the traditional case study methods to suit your need for technical content. In the past, books on case study described the case studies and asked the students to analyze them. Frequently, they failed to provide the technical knowledge that was needed to work on the case study. We have rectified this by adding chapters that describe the competency material required to analyze the case studies. We will explain these next.

Competency Material for Analyzing the Case Study

For the case study, we have created competency material that will help you learn the basic IT and business facts that are essential in order to analyze the case study. In addition, we refer you to library resources so that you can perform additional research. The competency material should give you a good overview of the industry and help you solve the problem mentioned in the case study.

The case studies included in this book provide a complete and detailed presentation of the subject under investigation (Zonabend, 1992), but they do not detail all the technologies used in a company. From this viewpoint, the case study only permits the understanding of a single facet that is intrinsic to the case under investigation (Gragg, 1954). However, a case study analysis gives you a detailed understanding of a particular technical problem that happened at a company.

The case studies presented in this textbook are research based and:

1. Are based on real-world empirical research since the authors worked with the managers and IT personnel, and went to the workplace to understand the problem, spending a considerable amount of time there during the actual decision and implementation of the option chosen by management.
2. Tell a story that has a beginning, background, and a conclusion.
3. Portray the situations accurately and report the reality of what happened in an unbiased and non-judgmental manner.

4. Have a decision focus, so that you have to consider alternatives and make decisions.
5. Are self-contained so that you have all the information needed to identify alternatives and make decisions.
6. Have identifiable characters, so that role playing is possible.
7. Include choices that are based on conflicting evidences.
8. Generate debate among the team members, since each case has more than one viable choice.
9. Include descriptions of the criteria used to evaluate the options.
10. Emphasize technical issues by including detailed technical information.
11. Emphasize business issues by including relevant information.

Why are Cases Used?

Cases enable you to learn by doing and by teaching others. What you learn thus becomes deeply ingrained and stays with you. The repetitive opportunity to identify, analyze and solve a number of issues in a variety of settings prepares you to become truly professional in your field of work.

The whole point of using cases is to allow you to take on the roles and responsibilities of specific people in specific organizations. Think of it as a form of on-the-job training. Cases provide an opportunity to become deeply involved in decisions actually faced by real people in real organizations; to take ownership, to feel the pressure, to recognize the risks, and to expose your ideas to others. Please recognize that the real life people whose roles you assume in cases were under pressure, may have felt butterflies in their stomachs and, in some instances, put their careers on the line.

Just imagine the experience you will accumulate over time while addressing these issues and challenges across a wide range of functional areas, levels of responsibilities, and types and sizes of organizations and industries, as well as locations throughout the world. You unlock a wealth of learning opportunities each time you put yourself into the decision maker's position. It is the cumulative impact of these different case challenges that will permit you to take on tasks in the future, knowing that the process of implementing ERP systems effectively has become a major personal asset.

Cases give you the chance to practice the art as well as the science of management in a laboratory setting, with little corporate and personal risk involved. In essence, cases are to management students what cadavers are to medical students, the opportunity to practice on the real thing harmlessly.

Cases are also an excellent tool to test the understanding of theory, to connect theory with application, and to develop theoretical insights. Cases provide the opportunity to see how theory applies in practice. Moreover, cases provide information about how work is planned and organized in various settings, how systems operate and how organizations compete.

Managers seldom have access to all the pertinent information when making decisions. Likewise, cases seldom contain all the information you would like. Thus, cases force you to make decisions with the available information, thereby helping you to tolerate incomplete information and ambiguity.

Because of the discussion-based format of the case method, cases are an excellent vehicle for developing your self-confidence, your ability to think independently and to work cooperatively. Moreover, cases foster the development of insights into your own strengths and weaknesses and allow for profound personal growth.

The case method of instruction is particularly well suited to dealing with new and complex situations. In the business world in particular, managers need to adapt to ever changing circumstances. Cases force you to think for yourself and generate your own learning. Cases actually engage you in a process of learning how to learn. While every case presented in this book is different, it is the process of learning how to learn that is generalizable.

Learning with cases provides you with an invaluable opportunity to develop a wide range of very useful skills.

Inventory of Skills Developed by the Case Method

1. *Analytical Skills*: The case method enables you to develop qualitative and quantitative frameworks to analyze business situations, including problem identification skills, data handling skills, and critical thinking skills. You are forced to reason clearly and logically in sifting carefully through the data available.
2. *Decision Making Skills*: The case method pushes you, on the basis of your analytical work, to assess what can be done and to make decisions. You will learn to generate different alternatives, to select decision criteria, to evaluate alternatives, to choose the best one, to formulate congruent action and implementation plans.
3. *Application Skills*: Cases provide an opportunity for you to practice using the tools, techniques, and theories you have learned.
4. *Oral Communication Skills*: The case method provides ample opportunity not only to listen to your colleagues, but also to express yourself, construct arguments and convince them of your views. Thus, a whole set of speaking,

listening and debating skills are developed. In this exchange of ideas and arguments, you learn to think on your feet, consider others' viewpoints, and take and defend your own positions.

5. *Time Management Skills*: Under the heavy pressure of case preparation and the juggling of your various other responsibilities, you are forced to schedule educational activities carefully and manage time effectively.
6. *Interpersonal or Social Skills*: The case method, through small group and large group discussion, promotes learning how to deal with your peers. This learning includes conflict resolution skills and practicing the art of compromise. Because so much of your future work life will involve committees, task forces, boards or project teams, learning to work effectively in a group will differentiate you from other candidates in the job market.
7. *Creative Skills*: Because no two business situations are quite the same, the case method encourages looking for and finding solutions geared to the unique circumstances of each case. This method invites you also to use your imagination in problem solving, as there are normally multiple solutions to each case.
8. *Written Communication Skills*: Through regular and effective note-taking, case reports and case exams, you learn the skills associated with effective writing. The emphasis on writing skills varies depending on the program you are enrolled in, but often takes on a high priority in business programs, as it is a key factor of success in management.

Development of the Robert Bosch Case Study

We developed the Robert Bosch case study during 1999/2000 by working with the heads of the IT division at Robert Bosch GmbH and Robert Bosch US. Chapters VI and VII were reviewed by these managers, who confirmed their accuracy and signed off on the results. During 2004, we again met with the heads of the IT division at Robert Bosch GmbH and Robert Bosch US. Chapter IX resulted from this visit. Again, the material in this chapter has been verified by the IT managers.

Our goal in presenting the Robert Bosch case study is to create an awareness of the strengths and complexities of an ERP implementation. In many instances, you may have little background knowledge of the specific ERP systems discussed in the case study. This case is designed to show the value of information technology in today's business world. It is also designed to generate an awareness of the complexities surrounding a ERP system, and the ensuing struggles that some companies face in their efforts to utilize the system to its fullest ... and in some instances merely to get the system functional! By describing the experiences of Robert Bosch GmbH in implementing the SAP R/3 in three different chapters (Chapters VI, VII, and IX) for the periods 1991-1997, 1997-2000, and 2000-2004, we are giving you useful information about the changes and the issues the IT managers faced as they worked to develop a functioning system. Robert Bosch US spent $30 million to implement the SAP R/3 system and bring it into use during the time periods discussed in these three chapters, and expect to spend about one billion dollars for the overall implementation effort. We expect you to benefit significantly as you study the details of the SAP R/3 implementation at Robert Bosch and use this learning effectively as you face the challenge of ERP implementations in your own professional career in the future.

The case is also designed to show the importance of aligning corporate strategy and information technology strategy, so that they will work together in an efficient manner at all levels. This relates back to the firm's global presence. If the company operates strictly at a regional or national level, then there are fewer tiers that this alignment has to meet. However, since Robert Bosch GmbH operates at a multinational level, this alignment needs to be congruent in ALL of the countries that the firm has operations in.

While the different information technologies discussed in this book may sound daunting and vague, the reality is that the more you understand about the topic, the better prepared you will be for your future career. The excitement of using ERP systems is not reserved for CEOs, CIOs, or the financial department. Information technologies stand to revolutionize the ways in which managers make their decisions. Therefore, it is important that you have a good understanding of ERP systems and how to implement them in real-world companies.

Overview of the Book

Chapter II discusses the possible architectures for a multinational company and describes the organizational structure of Robert Bosch GmbH (RB). Chapter III provides a basic understanding of the technical architecture of an ERP system and provides details about major ERP vendors. Chapter IV discusses the need for IT systems to be aligned with business needs, change management issues, and describes the phases in the change management life cycle. Chapter V describes the database management systems fundamentals and discusses concepts such as normalization and single instance. It lists some of the technical issues that need to be considered in implementing ERP systems. These chapters lay the theoretical foundation that you will need to analyze the Robert Bosch case study.

Chapter VI discusses the multiple information systems that were used at Robert Bosch during 1991-1998 and the efforts to standardize them. Chapter VII provides details of the use of SAP R/3 at Robert Bosch and the implementation experiences at Robert Bosch US. It also gives an overview of the structures that were considered by Robert Bosch US to standardize their information systems implementation. You will find that the final system implemented, described at the end of Chapter VII, was actually a work-in-progress and serious issues later arose that were related to the implementation experience and to working as a multinational team. Both Chapters VI and VII have assignments for you to complete at the end of the chapters, and you and your classmates will be assigned the roles of the managers involved in the ERP system implementation. Please work first by yourself, then in small groups, and finally in large groups to analyze the information provided in these two chapters. After you have done so, you can read through Chapter VIII. In this chapter, we will analyze the information system implementation discussed in the earlier chapters and provide recommendations as to what the company might have done at that time. Chapter IX concludes the case study by giving details of the developments at Robert Bosch during 2001-2004 as they continued to implement the ERP system. This chapter discusses the management changes that occurred both at the corporate level and at the U.S. level, the structure chosen by Robert Bosch US, the consolidations that took place, and the issues that they had to deal with during 2004. These four chapters will provide you with a great deal of practical knowledge concerning the implementation experience of Robert Bosch during the period 1991 to 2004.

Chapter X analyzes the technical architecture of RB's IT systems, discusses the future directions of SAP R/3 such as Netweaver product, and shows how RB could enhance its IT implementations by using these future offerings from SAP. Chapter XI presents another example of the use of ERP systems in another multinational corporation, Sidler GmbH. This example provides you with further information on the challenges involved in implementing SAP R/3 systems in a company. Chapter XII summarizes the material covered in this book and derives lessons for IT managers and students who are planning to implement ERP systems in their own organizations. It concludes by showing how you can improve your analytical, decision making, application, communication, time management, social, and creative skills by working on the case study. The chapter also provides an analysis of the Robert Bosch case study using the change management life cycle concepts described in Chapter IV.

Your Responsibility in Analyzing the Case Studies

This is a different type of book from other textbooks. While you will gain a great deal of information by simply reading through the book by yourself, you will obtain significant additional value if you work in small teams and large groups to analyze the case study.

The educational challenge provided by this case study can be measured in three dimensions with three levels of difficulty (1 being the easiest, 2 moderate, and 3 the hardest) (Mauffette-Leenders, Erskine, & Leenders, 1997b; Liang, Xue, Boulton, & Byrd, 2004; Esteves-Souza & Pastor-Collado, 2000). The *analytical* dimension raises the question: "What is the case reader's task with respect to the key decision or issue of the case?" The Robert Bosch case addresses several of these types of questions. For example, at the end of Chapter VI, the case notes: "Don, the CIO of Robert Bosch US had to establish a strategy to make multiple systems work together in order to cope with a growing and changing business." At the end of Chapter VII, the case concludes: "To make this SAP R/3 implementation project successful, Don not only had to choose a structure from among the five alternatives, but also had to design the implementation process." Thus, this case belongs in the third degree of analytical difficulty. The second dimension in which the case study could be measured is the *conceptual* dimension. It asks the question: "What theories, concepts, or techniques might be useful in the understanding and/or resolution of this case situation?" Chapter II provides theories about the structure of multinational corporations, Chapter III details theories concerning the structuring and harmonizing of SAP systems, Chapter IV presents the change management life cycle, and Chapter V discusses database management systems. In addition, you can refer to the literature for ERP implementation theories. Based on these theories, you must choose an alternative structure for implementation at the end of Chapter VII. The extensive time and effort involved in reading these materials and analyzing them gives a difficulty score of 3 to the conceptual measure of this case study. The third dimension in which a case study could be measured is the *presentation* dimension, which asks: "What is really important and relevant information here and what is still missing?" The case studies presented in Chapters VI and VII have been edited and shortened so that relevant information is presented and extraneous information has been removed as far as possible. As a result, it should be measured as a 2 in the presentation dimension. Thus, the Robert Bosch case study presented in this book would be classed as a (3,3,2) case. In order to derive valuable lessons from this case study, you have to read and analyze the case study individually, in small groups, and in large groups. Such a process will help you discuss and understand the principles, processes, acronyms, and jargon associated with implementing an ERP system.

- *Individual Preparation*: To get the most from the use of the Robert Bosch case study, you must take an active role in the learning process. You need to read the case studies thoroughly, work through the analysis, question others' arguments, and take an active part in the discussion. Why do you need to do all this? Your instructor will not be present to help you once you graduate. Selecting a recommendation and presenting it to a critical audience will be excellent experience for you. You will also have an opportunity to bring your outside experiences to the classroom and apply them to the decision being made. We expect you to consult with other instructors in creating your recommendation. Don't feel constrained if you feel that there are insufficient facts in the case study for you to perform a good analysis. We have presented you with a real-life situation and limited information is one of the factors that every professional and manager must learn to deal with in their jobs. Your task is to make sense of the information provided and use it to create viable recommendations. In order to analyze the case study, you must first understand the problem. Then you will be able to construct a range of feasible alternatives for solving the problem. The third step is to identify any criteria that may affect the decision. Finally, you must perform an analysis of the problem and choose the most appropriate solution.

- *Small Group Discussion*: In order to analyze the possible solutions to the assignments provided at the end of Chapters VI and VII, it will be beneficial if your team meets together before class and discusses the possible solutions. These small group discussions help in the following ways: you teach others and learn from others, everybody is more prepared to perform the analysis, every member speaks about the case, you develop better communication skills, you build relationships with others, and the meetings are likely to bring forth a host of new ideas that would have taken a lot longer for you to uncover alone. The final step is for your team to agree on your joint decision and create a written report and a presentation recommending it. When you prepare this report, take into account that other teams are going to be recommending alternate solutions. You can use the technical and business data provided in the chapters to strengthen your recommendation. We have also provided references to other books and Web sites so that you can perform further research and obtain other relevant information.
- *Large Group Discussion*: Your small group will make a presentation in the class. The purpose of the class is not simply to reach a thorough understanding of the case and an optimal resolution of the issue or decision. The class discussion will give you the opportunity to practice working in a large group, to share your individual and small group learning with others, and to be evaluated by peers and instructors. It is possible that your instructor might request the small groups to work together and come up with a solution that is acceptable to all groups. This type of presentation in the class is an excellent learning experience since you learn by doing, teach others, practice public speaking, test ideas, get good grades, and feel very

Figure 2. Learning with the three stages of learning

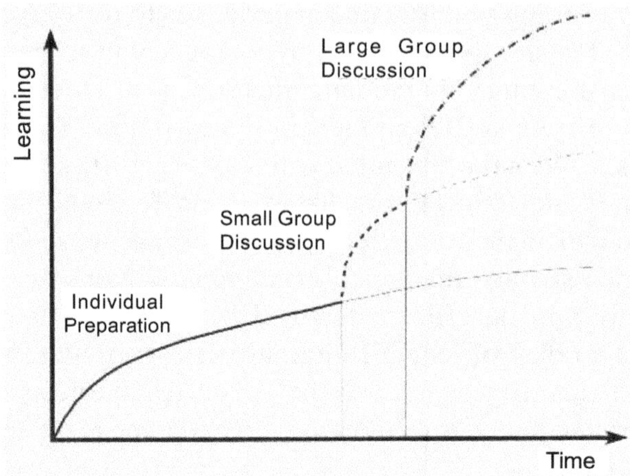

good about learning the complexities of ERP implementation experiences. Figure 2 shows the learning that could be achieved by following the three-stage process in case analysis. Each of these three stages is vital to effective learning and each contributes in different ways to maximizing the quantity and quality of learning. The figure shows how each of the three stages contributes to the quality and quantity of your learning in a progressive and cumulative fashion. The aim is to help you achieve the best possible results in a minimum amount of time. The combination of all three stages, executed properly and in the correct sequence, assures effective learning.

- *Feedback*: After you have presented your analysis in the classroom, you can compare your analysis to that provided in Chapter VIII. In addition, you can read Chapter IX to obtain detailed information on what actually happened at the company during 2001-2004. Please remember that this case study is from the real-world; there is no certainty that the managers' decisions were the most appropriate and it often takes many years before the true effectiveness of a decision can be measured. We hope that you enjoy this experience and that it helps you to see the connections between the different theories and problems faced in a real-world work environment. We hope you learn important and valuable lessons about ERP implementation experiences that you can use in your future career as a result of using this book.

Chapter Summary

Organizations are becoming bigger in size and global in operations and ERP systems are an essential part in standardizing the operations of the organization so that the transactions can be transparent to all members of the organization. Though relatively unimportant 30 years ago, a basic knowledge of computer operations and the Internet has now become essential for most employees. Similarly, even though knowledge of ERP systems is currently rather limited in most companies, in the near future learning about and using ERP systems will become essential. Therefore, as a future business leader, it is essential for you to obtain a good overview of ERP systems and how to implement them in companies.

This chapter discussed the need to learn about the implementation issues for ERP systems and then justified why case studies are an ideal way to learn about such issues. It went on to provide an overview of the 12 chapters in this book, along with guidelines on how you can learn valuable lessons by reading through the chapters both by yourself and working in teams.

Short Essay Questions

1. What are the key management issues facing information technology professionals? Discuss the importance of these issues to your future career.
2. Provide an overview of an ERP system.
3. What are the benefits of implementing an ERP system?
4. Define a case study.
5. Have you ever used a case study during your education? What are the advantages and disadvantages of using this instructional methodology?
6. Why will ERP systems play a major role in the future of business operations? Discuss.
7. Explain the steps that you could use to obtain the maximum benefit in analyzing the Robert Bosch GmbH SAP R/3 implementation case study.

References

Bancroft, N. H., Seip, H. & Sprengei, A. (1997). *Implementing SAP R/3: How to introduce a large system into a large organization.* Upper Saddle River, NJ: Pearson Prentice Hall.

Bonwell, C. C. & Eison, J. A. (1991). *Active learning: Creating excitement in the classroom.* Washington, DC: George Washington University.

Copeland, M. T. (1954). The genesis of case method in business instruction. In M. P. McNair (Ed.), *The case method at Harvard Business School.* New York: McGraw-Hill.

Daniels, M., Faulkner, X., & Newman, I. (2002). Open ended group projects, motivating students, and preparing them for the real world. *Proceedings 15th Conference on Software Education and Training (CSEE&T 2002)*, Covington, KY (pp. 115-126).

Davenport, T. H. (2000). *Mission critical: Realizing the promise of enterprise systems.* Boston: Harvard Business School Press.

DeFelice, A. (2005). *SMBs' rising backbone: The global ERP market.* Retrieved September 8, 2005, from http://www.destinationcrm.com/articles/default.asp?ArticleID=5220

Esteves-Souza, J., & Pastor-Collado, J. A. (2000). *Towards the unification of critical success factors for ERP implementations.* 10th Annual BIT Conference, Manchester, UK. Retrieved September 8, 2005, from http://www.army.mil/aeioo/docs/BIT2000.pdf

Fox, P. (2004, June 21). Big isn't better. *Computerworld, 38*(25), 20.

Frank, R., & Sidel, R. (2002, June 6). Firms that lived by the deal in '90s now sink by the dozen. *Wall Street Journal.*

Garcia, F., & Moreno, M. (2004). Software modeling techniques for a first course in software engineering: A workshop based approach. *IEEE Transactions on Education, 47*(2), 180-187.

Goodwin, B. (2005, April 4). Shortage of ERP skills could push up pay. *Computer Weekly.* Retrieved September 8, 2005, from http://www.computerweekly.com/Articles/2005/04/04/209199/ShortageofERPskillscouldpushuppay.htm#ContentContinues

Gragg, C. I. (1954). Because wisdom can't be told. In M. P. McNair (Ed.), *The case method at Harvard Business School* (p. 6). New York: McGraw-Hill.

Hamel, J., Dufour, S., & Fortin, D. (1993). *Case study methods.* Thousand Oaks, CA: Sage Publications.

Hitt, L., Wu, D. J., & Zhou, X. (2002). Investment in enterprise resource planning: Business impact and productivity measures. *Journal of MIS, 19*(1), 71-98.

Kale, V. (2000). *Implementing SAP R/3*. Indianapolis, IN: Sams Publishing.

Kletz, T. A. (1970). Putting knowledge to use. *Education in Chemistry, 7*(6), 229.

Krone, J., Juedes, D., & Sitharam, M. (2002). When theory meets practice: Enriching the CS curriculum through industrial case studies. *Proceedings 15th Conference on Software Education and Training (CSEE&T 2002)*, Covington, KY (pp. 207-214).

Lammers, W., & Murray, J. (2002). A profile of teaching techniques used in the university classroom. *Active Learning in Higher Education, 3*(1), 54-67.

Lamnias, C. (2002). The contemporary pedagogic device: Functional impositions and limitations. *Pedagogy, Culture and Society, 10*(1), 21-38.

Langley, N. (2003, April 22). Demand for SAP skills on the rise: Industry overview. *Computer Weekly*, p. 60.

Larocca, D. (1999). *Sams teach yourself SAP R/3 in 24 hours*. Indianapolis, IN: Sams Publishing.

Laudon, K. C., & Laudon, J. P. (2004). *Management information systems: Managing the digital firm*. Upper Saddle River, NJ: Prentice Hall.

Liang, H., Xue, Y., Boulton, W. R., & Byrd, T. A. (2004). Why western vendors don't dominate China's ERP market. *Communications of the ACM, 47*(7), 69-72.

Love, K. (2000). The regulation of argumentative reasoning in pedagogic discourse. *Discourse Studies, 2*(4), 420-451.

Mauffette-Leenders, L. A., Erskine, J. A., & Leenders, M. R. (1997a). *Learning with cases*. Ontario, Canada: Richard Ivey School of Business.

Mauffette-Leenders, L. A., Erskine, J. A., & Leenders, M. R. (1997b). Because wisdom can't be told. In M. P. McNair (Ed.), *The case method at Harvard Business School* (p. 6). New York: McGraw-Hill.

McDermott, B. (2004, June 14). CRN Interview—Bill McDermott, SAP America, Overview. *Computer Reseller News*, p. 29.

Naumes, W., & Naumes, M. J. (1999). *The art and craft of case writing*. Thousand Oaks, CA: Sage Publications.

North American Case Research Association (1997). *Guidelines for track chairs*.

Richardson, V. (2003). Constructivist pedagogy. *Teachers College Record, 105*(9), 1623-1640.

Robey, D., Ross, J. W., & Boudreau, M. C. (2002). Learning to implement enterprise systems: An exploratory study of the dialectics of change. *Journal of MIS, 19*(1), 17-46.

Seller, D. C., Turnwald, G. H., & Bull, K. S. (1994). From teaching to learning: Part III. Lectures and approaches to active learning. *Journal of Veterinary Medical Education, 21*(1), 1-9.

Spizizen, G., & Hart, C. W. (1985). Active learning and the case method. *The Cornell Hotel and Restaurant Administration Quarterly, 26*, 63-66.

TechWeb.com (2004, May 17). *ERP market benefiting from IT spending increase*. TechWeb News. Retrieved September 8, 2005, from www.itnews.com.au

Thorpe, M. (2000). Encouraging students to reflect as part of the assignment process. *Active Learning in Higher Education, 1*(1), 79-92.

Ul-Haq, R., Stiles, J., & Pond, K. (2003). Learning expectations and learning process design. *Active Learning in Higher Education, 4*(2), 168-180.

Welti, N. (1999). *Successful SAP R/3 implementation: Practical management of ERP projects*. New York: Addison-Wesley.

Zonabend, F. (1992). The monograph in European ethnology. *Current Sociology, 40*(1), 49-54.

Endnote

[1] For example, billions of dollars had to be spent to fix the Y2K bug and avert a major disaster, but how did the problem originate? For technical reasons, programmers in the 1970 and 1980s wrote computer code that only recognized two letters for the year field. The concern was that when the year rolled over to 2000, computers would read this as the year 1900, creating major problems. This problem would never have arisen if the programmers had discussed the requirements of systems with commercial users and asked if the code they wrote was likely to still be in use in the year 2000.

Chapter II

Multinational Companies

Learning Outcomes

- Understand different organization structure of multinational corporations
- Apply the organization structure models to analyze the global organization structure of Robert Bosch GmbH
- Analyze the regional organization structure for Robert Bosch US
- Obtain an overview of the theoretical models of business strategies used by multinational firms

History

The concept of multinational sales was not a viable option until modern transportation and communication systems were developed to a level that made dealing with large distances possible. Although in many parts of Europe companies were able to sell goods and services in their neighboring countries, this was only because the size of the countries was small enough for this to occur. In reality, it was not until the advent of transport trucks, container ships, cargo planes, and/or railway networks that the movement of products over large distances became feasible. The trend towards the internationalization of production in its broadest sense can be roughly divided into three stages; The first of these, during which direct foreign investment occurred on a considerable scale, began approximately in the middle of the 19th century and ended with the outbreak of World War I. The second stage, marked initially by a recovery of foreign activities by private enterprises but followed by stagnation and then an absolute decline in these activities, comprised the inter-war period. The third stage in this development started at the end of World War II and has been marked by an almost steady, large-scale growth of global business dealings, foreign direct investment in other countries, and direct cooperation among countries (European Union, 2005; OECD, 1977). The European Union now consists of 25 countries and covers a large part of the continent of Europe. Though richly diverse, the 25 countries that make up the EU are all committed to the same fundamental values: peace, democracy, the rule of law and respect for human rights. Trade between EU countries has become much easier and cheaper thanks to the removal of tariff barriers and custom duties. In such an environment, large corporations spread around the world have to continually recalculate where it is cheapest to produce quality products while simultaneously ensuring that the products are available for sale in the countries where their customers are located and when the products are needed and at appropriate prices so as to ensure profits.

What Multinational Companies Do

The goal of a company is to increase the wealth of its shareholders and benefit its employees and a primary method of doing this is by maintaining and/or increasing its ability to grow. Once a company has operated in a specific region or country for a long period of time, it can start to face problems with meeting its growth targets. One way to continue to grow is to expand into new and emerging markets, which entails branching out into the international marketplace. Consider, for example, a multinational manufacturing company that does business in a number of countries: it has a substantial commitment of its resources in international business, it engages in international production in a number of countries, and it has a worldwide perspective in its management. By branching out into multiple countries, the company continues to grow and reduces the risks it faces if conditions in one country changes drastically.

Organizational Structure

The multinational enterprise is generally structured in a similar fashion to a traditional national or regional company, although there are certain areas that are different and thus worth examining further. Many multinational companies utilize a standard structure throughout their sub-companies, but due to the sheer size of their operations they are often forced to make certain adjustments to their corporate infrastructure and to their business practices.

A multinational company has to blend three basic sets of components — functionality, product, and geography — in its organizational structure. The functional components include marketing, finance, production, logistics, research and development, personnel, and others. The product components involve matters such as how best to design, adapt, produce, distribute, sell, promote, and service specific product lines. The geographic components involve a detailed knowledge of national and regional environments and how best to undertake the company's business in each. While other considerations are also involved, multinational companies tend to develop their organizational structure in response to the degree of their international involvement (Barlett, Ghoasl, & Birkinshaw, 2003; Dymsza, 1972).

A multinational corporation might base its corporate structure on one of five models as follows: Functional Structure, Product Structure, Geographical Structure, Matrix Structure and Network Structure (Katsioloudes, 2002). While none of these models can be termed "the best", it may well be that the most appropriate structure is one that combines elements of some or all of these models. The possible structures of the multinational organizations are described below:

A. Functional Structure

Perhaps the most traditional of the five, the *Functional Structure* is organized around common organizational functions such as marketing, production, research and development and accounting, and is probably the most commonly used structure. This structure is shown in its base form in Figure 1.

B. Product Structure

This form of organizational structure is characterized by the corporation revolving around its organizational outputs (i.e., products), and it tends to be utilized in environments that experience moderate to high uncertainty and change. This structure is shown in its base form in Figure 2.

Figure 1. Functional organizational structure

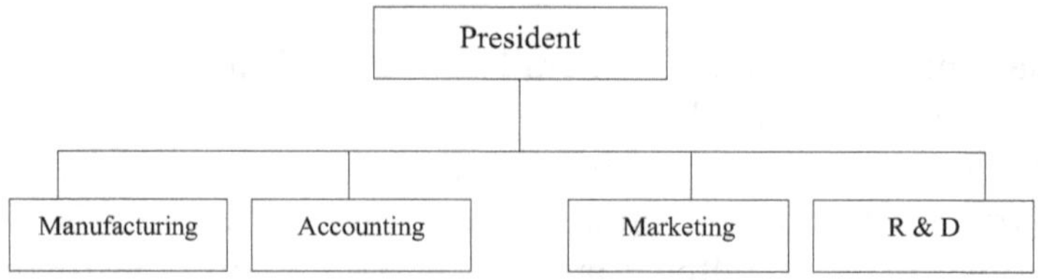

Figure 2. Product organizational structure

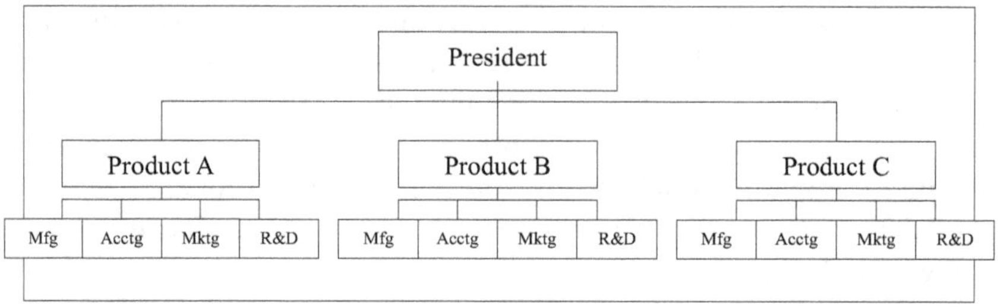

Source: Katbsioloudes, 2002

C. Geographical Structure

While similar to the Product Structure, the *Geographical Structure* approach requires the organization to develop specific geographical regions of their market, where each region includes the functions necessary to produce and market the product(s). The fundamental difference between this structure and the Product Structure is that instead of revolving around products, it revolves around its regional consumers. This sort of structure tends to be utilized when the consumers in each region have specific tastes and preferences and, therefore, requires that each regional unit has a certain amount of regional expertise. The structure is shown in its base form in Figure 3.

D. Matrix Structure

As mentioned previously, many corporations choose to incorporate different facets from different models into their corporate structure. Structures that are

Figure 3. Geographical organizational structure

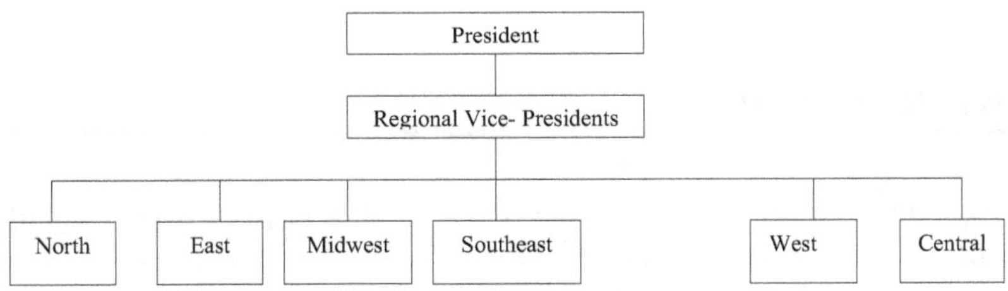

Figure 4. Matrix organizational structure

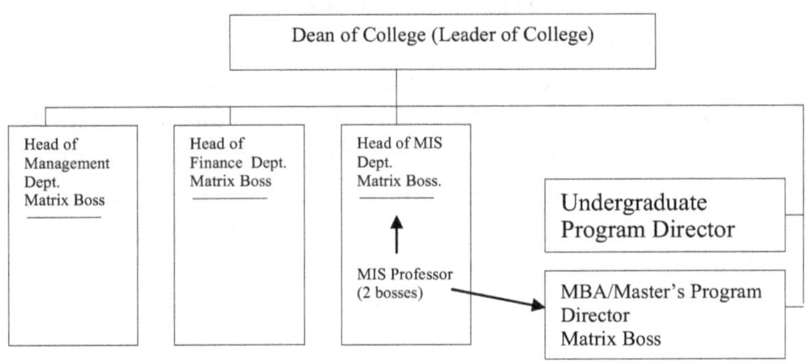

derived from different models are referred to as *Matrix Structures*. In today's complex business environment, many corporations' structures have become just as complex. In many cases, these complex hierarchies are known as Matrix Structures. Perhaps one of the key identifiers of a matrix-type structure is the presence of dual, or multiple, reporting relationships. By comparison, in the previous examples the hierarchy flows directly up and down, with each level overseeing only the level directly below it and reporting only to the level directly above it, but a matrix structure can have multiple "bosses". However, not every level of the matrix is in a position of reporting to two or more superiors and in many cases, most employees will report directly to only one superior, although the presence of some employees who report to multiple superiors is indicative of a matrix arrangement. A matrix structure for a typical university college of business is shown in Figure 4 to demonstrate how this might occur. This example shows how a faculty member might be responsible to two people: the head of his or her department and the head of the college's MBA/master's degree program. Assuming this faculty member teaches mostly graduate level courses, although formal evaluation of the faculty member might be done by the department head, the teaching assignments and feedback will come from the MBA/master's degree program director.

E. Network Structure

In some ways the *Network Structure* is not a structure in precisely the same way as the models previously discussed. The basis for this structure is an organization built of alliances with other organizations. In today's economy, outsourcing and downsizing are becoming important strategic initiatives and this structure allows for maximum flexibility and lean administration. A base schematic is provided in Figure 5.

Figure 6 shows a possible organizational structure for a multinational corporation that incorporates various dimensions from some of the previous models. The description takes you through the levels of reporting in this structure and demonstrates the flow of information through the example. The figure is a generalized corporate structure of a four-nation multinational enterprise. The first box labeled "Head Office" represents the company's global headquarters, where the Chief Executive Officer (CEO), Chief Financial Officer (CFO) and the Board of Directors are typically based. It is the pinnacle of the power structure and depending on the corporate culture, the vast majority of decisions regarding the direction of the firm are developed here. The next tier in the diagram has boxes labeled "N.O.1" through "N.O.4". These boxes represent the national head offices for the corporation. These national offices normally have a miniaturized version of the corporation's global power structure, although the

Figure 5. Network organizational structure

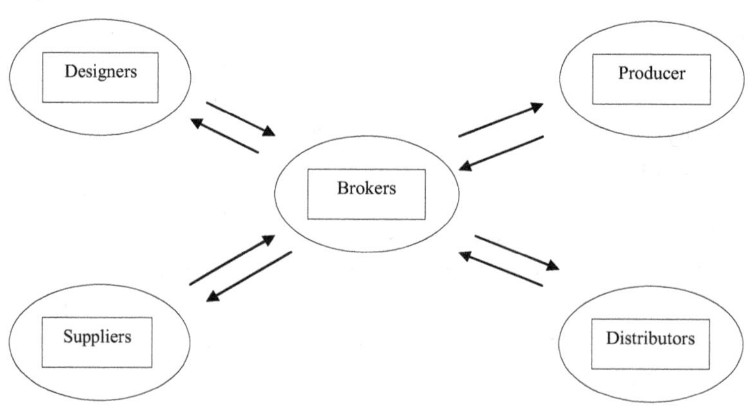

Figure 6. A possible structure of a multinational corporation

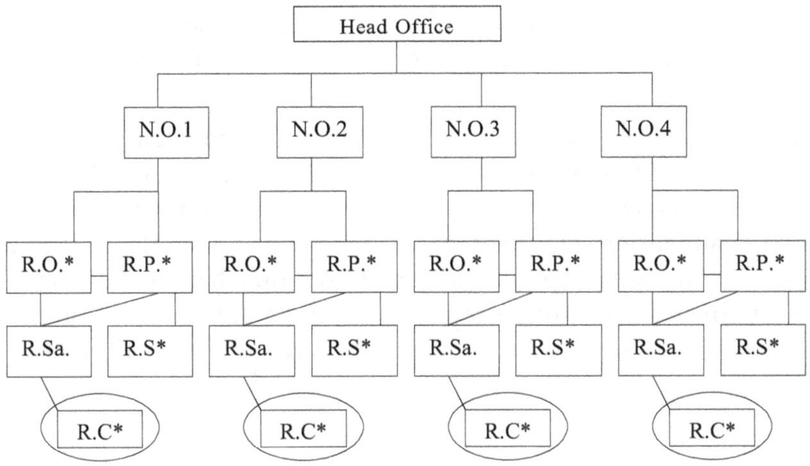

title usually given to the head of a national office is "President – Country". While each multinational has its own way of denoting power through its organizational structure, it will help to think of the head of the national entity as a president.

The next tier is labeled "R.O.*" and "R.P.*". These represent Regional Offices and Regional Production, respectively. The "*" indicates that there may be more than one of these in each nation. A Vice President of Operations typically runs these regional offices and a Vice President of Production runs the regional production facilities. Again, there are many possible variations in these titles, based on the corporate culture of the individual firm, but generally speaking, the regional manager is thought of as a Vice President.

The next tier is labeled "R.Sa" and "R.S*". "R.Sa." denotes the regional sales force and "R.S*" denotes the regional suppliers. The final tier is labeled "R.C*", which represents the regional consumer. Again, the "*" signifies that multiple suppliers and consumers are likely to be present.

While nearly all corporations are organized in a hierarchical fashion, the depth of the hierarchies plays an important role in day-to-day operations. Where a small organization is capable of responding to changes in its business environment very quickly, the sheer size of a multinational, with its deep hierarchy, dictates a slightly different approach. The majority, if not all, of the day-to-day decision making power is transferred from the global head office to the respective national head offices. This allows each country to run its operations as a national corporation and to manage its business in a timely fashion. This effectively removes some of the control previously held by the upper tiers of the managerial structure over the normal operating procedures, reserving their decision-making for larger issues. In the flat hierarchy of a small and/or national or regional enterprise, the upper levels of management are not as far removed from the bottom tier, and are therefore more effective at being a part of that process.

The market for the multinational corporation can be seen as the regional consumer. The multinational organization organizes itself so that it can sell its products to regional consumers in the most effective manner possible. While a company's headquarters may be located in Zurich, Switzerland, it must efficiently produce and sell its products in every region where it is has a presence. In order to illustrate the organizational structure of a large multinational company, we will examine the structure of Robert Bosch GmbH in more detail.

Organizational Structure of Robert Bosch GmbH[ii] (RB)

Robert Bosch, or Bosch, is known throughout the world and is synonymous with high quality products such as electrical and electronic automobile equipment, communications technology, power tools and household appliances. If you open the hood of your car, it is possible that you will find one or more products manufactured by Bosch. Founded in 1886 as the "Workshop for Precision Mechanics and Electrical Engineering" in Stuttgart, Germany, Robert Bosch soon began to specialize in ignition systems for motor vehicles and engines of all kinds.

Today, Robert Bosch GmbH differs from many global companies not only in its diversity and breadth, but also in its corporate structure and mission. While most companies are expected to grow and return profits to benefit their shareholders, RB is 92% owned by an institution, the Robert Bosch Foundation. This non-profit foundation is one of the largest of its kind in Germany and its existence as majority shareholder at Bosch, is perhaps the key difference between RB and other, more typical multinationals. The profits that the Robert Bosch Foundation receives from RB are used to fund healthcare, international understanding, welfare, culture, education, and other philanthropic ventures. This means that there is less pressure on RB to return immediate profits to its shareholders, at the expense of quality control, innovation and other costly endeavors. Since the profits do not directly increase the wealth of the shareholders but go to improve the society, Robert Bosch and its subsidiaries are given considerable leeway to innovate.

Bosch employs approximately 238,000 people, more than half of whom are employed in facilities outside of Germany. osch's global sales of roughly 40 billion Euros, are generated in nearly 250 subsidiaries and associated companies in more than 50 countries. The first international branches were set up at the end of the 19th century. Bosch operates roughly 260 production sites worldwide, of which nearly 200 are located outside Germany — in Europe, North and South America, Africa, Asia and Australia. Seventy two percent of the sales are accounted for by international business.

Robert Bosch's main business sector is Automotive Technology, which accounted for roughly 63% of their revenues and about 25.3 billion Euros during 2004. There are 10 product divisions, denoted K1 through K9 and KH[iii]. There are two other business sectors, namely Consumer Goods and Building Technology, accounting for 9.5 billion Euros in sales, and Industrial Technology, accounting for 5.2 billion Euros in sales during 2004.

The head office and board of directors for Bosch reside in Germany, and this by itself presents obstacles to managers, as not only are production facilities in geographically distant locations around the world, they also operate in different time zones and different business climates. Figure 7 depicts a simplified structure of Robert Bosch GmbH.

This figure shows the division of power among the executives at Robert Bosch. While the diagram does not show the complete reporting chain for all divisions, it does give a sense of how complex the management structure is. This figure is further condensed and inserted into the 'Executive' tier in Figure 8.

As this figure suggests, the corporate structure of Robert Bosch is extremely complex. However, what the figure fails to convey is the true complexity and depth of the hierarchy. The figure only shows the K1, K2,... divisions in the automotive sector and, as was alluded to earlier, the other two sectors also have multiple divisions. In addition, as previously mentioned, the Robert Bosch group operates in 50 countries, and in these countries, Bosch operates different divisions. K1 for example, deals with Anti-lock Breaking Systems (ABS). If Bosch manufactures some or all of its ABS lines in every country, that would necessitate 50 boxes below K1 in the hierarchy. These 50 boxes would merely represent the division heads for each country, and manufacturing would need an entirely new tier in the hierarchy. Imagine this Web spread across all the divisions and you will see the true depth and complexity of a company such as this.

The divisional structure is dynamic and has changed slightly in the past few years and doubtless will continue to do so in the future. You can view the current divisional structure of Robert Bosch GmbH by visiting their Web site at www.bosch.com.

Figure 7. Simplified executive structure

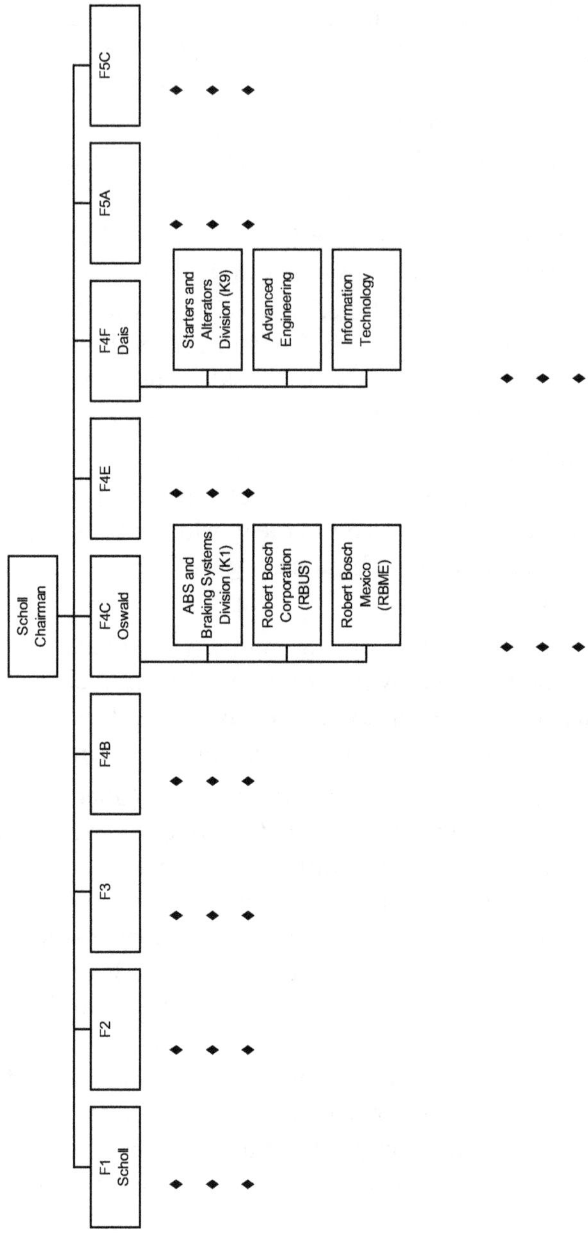

Figure 8. Executive tier at Robert Bosch GmbH

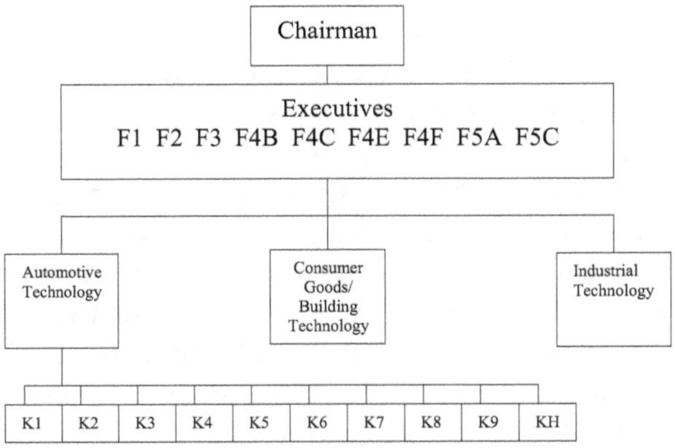

Multinational Strategies

In order for companies to function effectively in different countries, it is not sufficient to have effective structures alone; strategies need to be developed to manage them effectively. Although many strategies could be used, four major strategies that firms follow when they are involved in international business are listed and defined (Hill, 1998).

An *International Strategy* is where firms try to create value by transferring valuable skills and products to foreign markets where indigenous competitors lack those skills and products. This strategy makes sense if a firm has valuable core competencies that indigenous competitors in foreign markets lack and if the firm faces relatively weak pressures for local responsiveness.

A *Multidomestic Strategy* is when a firm tries to position itself to maximize its local responsiveness. This is similar to an International Strategy in the sense that it brings both skills and products to foreign markets, but it is characterized by the customization of its skills and products for each individual foreign market. While there are high costs involved in implementing this strategy, it is utilized when a market demands that a firm has a quick local response to changes.

In a *Global Strategy,* firms are characterized by offering standardized products to every market that they participate in. This allows them to take advantage of cost reductions from economies of scale and experience.

A *Transnational Strategy* is effectively a combination of the Global and Multidomestic Strategies. Its principal feature is that valuable core competencies do not exist in the home country alone, but in every market where the firm operates. This means that there is an important two-way flow of information, experience and expertise from the home market to the foreign markets and vice versa.

International firms can be further divided into two forms, *global industries* and *multidomestic* industries (Porter, 1986, pp. 9-40). At one end of the spectrum, global firms are characterized by the firm's competitive position in one market being significantly influenced by its position in every other market. At the other extreme, multidomestic companies are present in many countries, but their competition and competitive position differ on a country-by-country basis. The implications of a company's strategy lying between the two strategies is quite profound due to the structural and managerial implications of each choice. A multidomestic firm must manage its subsidiaries like a portfolio, in that each operation controls all of the important activities necessary to do business in that region, thereby giving them a high degree of autonomy. Global firms, on the other hand, seek to exploit competitive advantages by integrating activities on a worldwide basis in order to capture linkages between countries.

Robert Bosch GmbH is a good example of a firm that generally tries to follow a global strategy. While it can be argued that its individual business units operate with a good deal of autonomy and entrepreneurship, the key feature is that the business practices that make Bosch successful in Germany are the same principles that make it successful worldwide. Bosch manufactures automotive products that are more or less similar in the different countries where they are marketed. Because Bosch offers similar and/or identical products[iv] in its global markets it can, for example, exploit economies of scale and practice a centralized research and development strategy. These economies translate into a competitive cost advantage in each of its international markets. Thereby, Robert Bosch GmbH tends to adopt a global strategy in its operations.

As a global company, Bosch has to deal with the business climate of each nation it operates in. This presents a challenge in creating a unified business strategy. The business practices in Germany may be very different from those in the United States or in any other nation, and thus could lead to difficulties in creating and operating a global strategy.

Alignment of Corporate Strategies with Information Systems

A survey of 539 heads of IT from a broad range of industries, including manufacturing, government, health care, technology, education, and finance, by *CIO Magazine* (Cosgrove, 2003) showed that the most important issue facing the CIOs was to find best practices for the IT unit to partner with other business units and deliver the greatest value to the organization. The IT unit's greatest challenge was in prioritizing demands from various business units and aligning IT with business goals. The CIOs stated that in order to succeed in their jobs, effective communication and understanding business processes and operations was the most important. In addition, strategic thinking and planning has become pivotal in succeeding in the business climate. This means taking an enterprise wide view, rather than simply focusing on their department. It also means welcoming the business heads into the planning process rather than assuming that they will turn up providentially when the time comes to make critical IT decisions. Investing in systems and processes that enable strategic planning and the alignment of IS and business goals was also revealed to be a very important spending priority for most of the CIOs (Prewitt & Overby, 2003).

IT and business alignment tops the list of concerns and priorities for CIOs in 2004, according to a number of recent IT executive studies. As IT budgets start to finally open up again, business executives are counting on their CIOs to make smart investment decisions that positively affect the business bottom line. Business sense for a CIO these days is no longer just gravy — it's key to their survival ("Business / IT Alignment", 2004).

According to Joe Michini, CEO (former CIO), CandleTyme Stores (Jahnk, 2004), "the problem exists in one or all of three areas. First, many IT decisions are driven by business executives who know little about technology — except what they read in magazines or have been told by salespeople. This group includes CEOs, CFOs and COOs. They look to technology to drive the company. They believe that ERP, SCM, KM, data mining and a variety of software solutions will enhance revenue through efficiency gains or new customer sources. In some cases they do, but in most, the costs offset the gains. The company is structurally stronger, but on paper it looks the same or worse.

Second, many companies are directed by IT organizations that are technology-driven but don't understand the real needs of the business. They cannot translate business needs into technology solutions. Many IT executives cannot present a business case for or against a particular technology. Their condemnation or approval is based entirely on the performance of the technology. (Is it full of bugs or is it stable? Fast or slow? Easy to implement or hard? Can the vendor support it?) They present factors that should be considered during the technology evaluation stage but not at a strategic alignment level.

Third, those who run the business and those who run technology cannot agree on what alignment is. In reality, it is all perception based on expectations. There are companies that could be greatly enhanced through the use of technology but believe that they are perfectly aligned already. There are others that have the most beneficial technology available and don't know it.

In most cases, the solution is a strong CIO who knows both business and technology. An individual who performs a cost-benefit analysis first but also intimately understands the requirements of the technology. Someone who also has the authority to say no when other business units attempt to implement technology without the guidance or assistance of the IT organization. They are few and far between. "

According to Luftman (2000, 2003), "A decade of research has found that the key is building the right relationships and processes, and providing necessary training. A new tool, modeled after the well-known Capability Maturity Model (CMM) developed by Carnegie Mellon's Software Engineering Institute, can help you assess whether your company is on the path to alignment. Developers, businesspeople, and end users often have competing and even incompatible needs, but balance can be achieved with this method.

In business-IT alignment, IT is applied in an appropriate and timely way, in harmony and collaboration with business needs, goals, and strategies. Terms such as "harmony," "linkage," "fusion," "fit," "match," and "integration" are often used synonymously with the term "alignment." What follows is a methodology for assessing your company's alignment status. The CMM model it's patterned after has proven to be a powerful tool for managing development projects; however, until now, it focused on technology, rather than strategic business practices.

This alignment-assessment tool was successfully tested at more than 60 Global 2000 companies and is the subject of a benchmarking study jointly sponsored by SIM and The Conference Board. It has been applied in large and small companies, at all levels. The alignment model lets you measure how well the technical and business organizations work together, or it can be used to assess alignment within the IT department itself. It examines six dimensions, rating each on a scale of 1 (lowest) to 5 (highest).

- *Communications Maturity*: How well do the technical and business folks understand each other? Do they connect easily and frequently? Does your company communicate effectively with consultants, vendors, and partners? Does it disseminate organizational learning internally?
- *Competency/Value-Measurement Maturity* How well does your company measure its own performance and the value of its projects? After projects are completed, do you evaluate what went right and what went wrong? Do you improve your internal processes so the next project will be better?
- *Governance Maturity*: This dimension deals with how well the company connects its business strategy to IT priorities, technical planning, and

budgeting. Do the projects you undertake flow from an understanding of the business strategy? Do they support that strategy? If not, there may be conflict between the technical and business organizations. Simply put, IT governance is about who makes the decisions (power), why they make them (alignment), and how they make them (decision process).

- *Partnership Maturity*: To what extent have business and IT departments forged true partnerships based on mutual trust and sharing risks and rewards?
- *Scope and Architecture Maturity*: To what extent has technology evolved to become more than just business support? How has it helped the business to grow, compete, and profit?
- *Skills Maturity*: Does the staff have the skills needed to be effective? How well does the technical staff understand business drivers and speak the language of business? How well does the business staff understand relevant technology concepts?

To conduct an assessment of your company, the first step is to assemble a team of IT and business representatives to perform the assessment. For a single business unit, a team of 10 to 12 people would be typical. You would use fewer for a single department and more for a larger business unit. The team is responsible for assessing each of the six dimensions described above using a 1 to 5 scale, where:

1 = Doesn't fit the company, or the company is very ineffective

2 = Low level of fit for the company

3 = Moderate fit for the company, or the company is moderately effective

4 = Fits most of the company

5 = Strong level of fit throughout the company, or the company is very effective

It's important to have both business and technical people participate in evaluating each of the six dimensions. The goal is to come to an agreement on the level assigned. This may not be easy because the initial review will typically produce divergent results. When you find such divergence, you've identified a problem and an opportunity to improve the situation.

A trained facilitator may be valuable in keeping the discussion focused and honest. The team can use the information gleaned through the assessment process to create a plan for moving from the current level of alignment to the next level of maturity.

Experience shows that no single activity will enable a company to attain and sustain alignment. There are too many variables, and the technology and business environments are too dynamic. The strategic-alignment maturity assessment provides a vehicle to evaluate where a company is and where it needs to go to achieve and build business-IT alignment. The careful assessment of a company's alignment maturity is an important step in identifying the specific actions necessary to ensure that IT is used appropriately. The journey continues. The time frame depends on you."

Chapter Summary

This chapter provides an overview of the types of organizational structures used by multinational organizations so that you can understand and relate to the organizational structure that you might find in your company/ organization. The chapter then provides an example of the organizational structure of Robert Bosch GmbH. The chapter concludes by discussing the different strategies that are used by multinational corporations in managing their operations and shows the importance of aligning IT strategies with that of the organization.

Short Essay Questions

1. Briefly outline the three stages of internationalization.
2. Provide a few reasons why companies expand beyond a country's borders.
3. Look at popular business magazines such as Businessweek, Fortune, or Forbes. Identify a multinational corporation and discuss its organizational structure.
4. List and sketch the five organizational structures.
5. Which models, or combination of models, best matches Robert Bosch GmbH as described in this chapter?
6. List and describe the four multinational strategies. Which strategy best describes Robert Bosch GmbH?
7. Using additional resources (Internet, etc.) determine if the Robert Bosch GmbH corporate structure has evolved since this chapter was written. If so, how? Include a diagram.

8. Using additional library resources, identify another multinational corporation. What is its corporate structure? What is its business strategy?

References

Barlett, C. A., Ghoasl, S., & Birkinshaw, J. (2003). *Transnational management: Text and cases*. New York: McGraw-Hill.

Cosgrove, L. (2003, April 1). Survey results: The state of the CIO. *CIO Magazine*. Retrieved September 4, 2005, from http://www.cio.com/archive/040103/index.html

Dymsza, W. A. (1972). *Multinational business strategy*. New York: McGraw-Hill Publishers.

European Union (2005). *Europa, gateway to the European Union.* Retrieved September 7, 2005, from europa.eu.int

Hill, C. W. L. (1998). *International business: Competing in the global marketplace.* New York: Irwin/McGraw-Hill.

Jahnk, A. (2004, June 1). Why is business-IT alignment so difficult. *CIO Magazine*. Retrieved September 7, 2005, from http://www.cio.com/archive/060104/soundoff.html

Luftman, J. (2000, December). Assessing business-IT alignment maturity. *Communications of the Association of Information Systems 4,* Article 14. Retrieved September 7, 2005, from http://cais.isworld.org/articles/4-14/default.asp?View=html&x=32&y=11

Luftman, J. (2003, December). Measure your business-IT alignment. *Optimize Issue 26*. Retrieved September 7, 2005, from http://www.optimizemag.com/article/showArticle.jhtml?articleId=17701026&pgno=1

Katsioloudes, M. (2002). *Global strategic planning: Cultural perspectives for profit and nonprofit organizations*. Boston: Butterworth-Heinemann.

OECD (1977). Restrictive business practices of multinational enterprises. *OECD Report, RBP (77)/171-MNE*, Paris.

Prewitt, E. & Overby, S. (2003, April 1). The importance of being strategic. *CIO Magazine*. Retrieved September 4, 2005, from http://www.cio.com/archive/040103/index.html

Porter, M. E. (1986). Changing patterns of international competition. *California Management Review, 28*(2), 9-40.

SearchCio.com (2004, May 4). *Executive guide: Business / IT alignment.* Retrieved September 4, 2005, from http://searchcio.techtarget.com/general/0,295582,sid19_gci1049794,00.html

Endnotes

i The original version of this chapter was contributed by Keegan B. Corcoran, MBA.

ii GmbH is the German acronym for "Limited Corporation".

iii The acronyms K1, K2,... have changed. For current information, please check the Web site at www.bosch.com.

iv Basically, the products are similar or identical, although this does not take into account products that must be modified to comply with regional laws and policies.

Chapter III

Fundamentals of Enterprise Resource Planning (ERP)

Learning Outcomes

- Understand the interrelationship between MRP, MRP II, and ERP
- Be able to explain the components of an ERP system
- Have a basic understanding of the technical architecture of an ERP system
- Be familiar with the main players in the ERP market
- Know why SAP is a market leader

Historical Basis of ERP

Businesses first began to use IT technologies and computers on a routine basis in the 1950s and 1960s. The introduction of computers into the business environment required new systems that combined IT and business processes. In the early days, the most important systems in manufacturing companies were known as MRP (Material Requirements Planning) based systems. After two decades, MRP systems evolved into MRP II (Manufacturing Resource Planning), but it was many years before ERP (Enterprise Resource Planning) systems were first implemented, and these systems continue to evolve. To understand how today's ERP systems developed, let's first discuss MRP.

Material Requirements Planning

In the 1960s, Material Requirements Planning (MRP) emerged with the rapid evolution of computers. The main emphasis of these systems was to manage inventory, and the use of MRP helped companies control their inventory based on actual demand rather than reorder points. To do this, MRP used a set of techniques that took into account bills of material data, inventory data, and the master production schedule to predict future requirements for materials. The idea behind this was actually very simple. A finished product was subdivided into its components, and for every component a time schedule was developed. Based on this list, all necessary information required for the production of this specific product could be obtained in a very short time. The critical subcomponents could be tracked easily and, if necessary, could be obtained quickly to support on-time production. The critical time path could be defined and orders could be organized in order to prevent time delays in receipt of materials. However, even this simple procedure became tedious once the number of parts increased. Thus, a computer was essential to carry out these features of MRP. To sum up the benefits of MRP, it reduced the level of inventory a company needed to maintain, reduced production times by improving coordination and avoiding delays, and increased the company's overall efficiency.

Manufacturing Resource Planning

Unfortunately, with the simultaneous development of Information Systems, MRP was no longer sufficient to meet the new requirements of many companies. In the 1980s, companies transitioned to Manufacturing Resource Planning (MRP II). This system allowed manufacturers to optimize materials, procurement, manufacturing processes, and so forth, while at the same time providing financial and planning reports. The underlying idea behind the MRP II concept was to integrate MRP with further manufacturing functions and other business units. MRP II was designed to assist in the effective planning of all the resources available to a manufacturing company. Ideally, it addressed operational planning in units, financial planning in dollars, and included a simulation capability with which to answer "what if" questions. It included business planning, sales and operations planning, production scheduling, material requirements planning (as in the original MRP), and capacity requirements planning, along with executive support systems that could be used to balance capacities and materials (*The Free Dictionary*, 2005).

Integration posed a major challenge in the evolution from MRP to MRP II. Towards the end of the 1980s, many business processes such as logistics, procurement, and financial accounting needed to be integrated to allow companies to operate at their maximum efficiency. However, the issue was how best to do this.

Enterprise Resource Planning Systems (ERP)

ERP, or Enterprise Resource Planning Systems, have evolved considerably over 40 years as a result of continuous improvements in business management and the development of new information technologies. The main emphasis of ERP has been to integrate and synchronize the isolated functions into streamlined business processes (Tuteja, 2005).

The ERP concept was first implemented at the end of the 1980s with the development of better client/server technology that enabled the implementation of an ERP system. Client/Server technology will be explained later in this chapter. Figure 1 shows how the systems have expanded to encompass more areas of operations in a company.

Enterprise Resource Planning (ERP) is a cross-functional enterprise backbone that integrates and automates many internal business processes and information systems within the sales and distribution, production, logistics, accounting and human resource functions of a company (O'Brien, 2004, p. 194).

Actually, software systems to automate each of these internal business processes already existed and these were very efficient in their own areas. However, their relative autonomy and limited real-time interaction was a major problem that had to be solved. The divisions did not exchange data with each other, or even if they did exchange data it was poorly coordinated, which caused substantial problems that decreased the efficiency of the systems. For example, it was impossible for accounting systems to exchange data with manufacturing

Figure 1. Scope of MRP, MRP II, and ERP

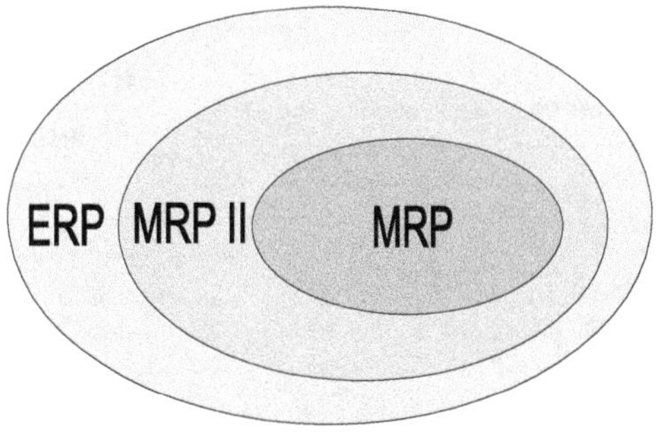

systems, and the time lag for exchanging data was so large that it brought no benefits for either division.

To illustrate this lack of coordination and inability to exchange information, an example will be helpful. The sales team may not know exactly what the production levels are, or what is actually in inventory. We say "actually" because even if the sales force has an idea of the inventory levels, once an order has been processed there is an immediate drop in actual inventory levels, as a time lag exists in the recalibration of the known inventory levels.

ERP not only coordinates several divisions but also enables companies to enter data only once for the information to be distributed to all the integrated business processes. ERP systems consist of several integrated suites of software modules which share common data and provide connectivity. Once the data has been recorded, it is available for all the company's divisions. The information about the processes in the company is represented consistently and is up to date in all business divisions at all times.

Figure 2 shows how the different divisions may be coordinated and integrated with an ERP system.

Logistics operations can issue a purchase order and enter into the system. That information is then immediately available for all other divisions. For example, when the financial department makes a payment, as soon as the payment is posted the relevant general ledger account will be updated. When the ordered material arrives it can either be sent to the warehouse or directly to the production process. As soon as it enters the warehouse, production can see that the material is in the warehouse and act accordingly. As can be seen, the integration potential is quite high even in this very simple example. It is important to understand how the system is supposed to work, before we discuss why it is so difficult to make it work seamlessly in a real organization. The above example

Figure 2. Before and after ERP (AEIOO, 2005)

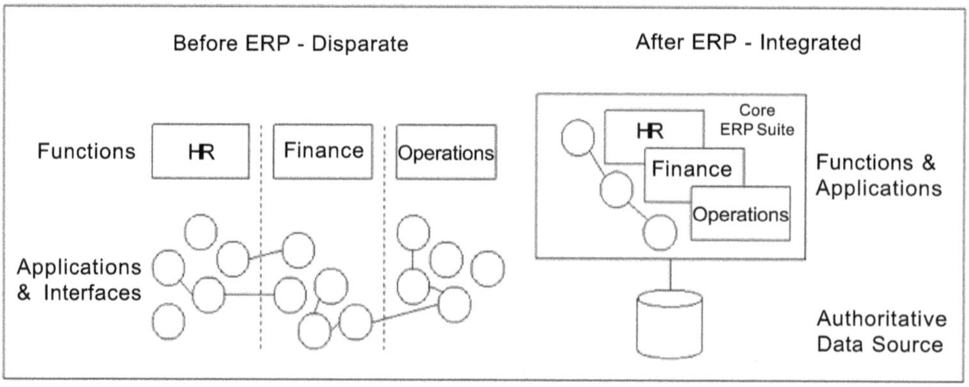

also shows which divisions might be included in the ERP system and how they might interact with each other. Each module (sales and distribution, accounting, human resources) will be discussed in detail in the next section and the interactions among these modules will be explained.

Components of an ERP System

Up to now, we have explained how ERP systems consist of several modules that support the business processes of a company. Now, the most important modules and their functionalities will be examined more closely. Many classification possibilities exist for ERP modules, but we shall classify the ERP systems into three main groups:

- Financials
- Logistics
- Human Resources

These main groups are each composed of several subfunctionalities, which will be outlined briefly under each module.

Financials

Financials, also known as accounting, are the processes involved in recording, summarizing, classifying, reporting, and analyzing the financial information of the business. These transactions have been more or less the same for decades. Unfortunately, current business processes can no longer be supported with traditional accounting systems alone, as changes in information technology and corporate restructuring have resulted in new accounting needs. Also, e-commerce, business partners, third-party producers, and outside vendors have placed new demands on these concepts. As a result, companies have to integrate their business processes with their accounting systems so that they can function more efficiently. With this integration, all parts of the company, from development to processes in logistics have access to information and accounting functions become enablers rather than simple "bean counters." Not only internal links are involved; important process links to and from vendors, customers, and financial institutions become possible with ERP-based accounting systems (Curran & Ladd, 2000).

Accounting can be divided into two groups: financial accounting and cost accounting. Financial accounting is responsible for interactions with outside vendors or suppliers, whereas cost accounting or managerial accounting is mainly responsible for a company's internal requirements, which are mainly used for controlling operations.

The accounting functions of an ERP system consist of several modules which are targeted to solve specific problems, such as accounts payable, accounts receivable, the general ledger, and billing and invoicing. We will describe each of these modules in turn below.

- *Accounts Payable*: The amount owed to suppliers, and the purchases from and the payments to suppliers are all recorded under accounts payable.
- *Accounts Receivable*: The amounts due from customers, along with the billing or invoicing of customers, is done in accounts receivable.
- *Billing and Invoicing*: Commercial documents issued by the seller to the buyer or vice versa, which include information about the products sold or bought, the quantity and the price, are included in billing and invoicing.
- *General Ledger*: Information from other accounting modules is consolidated and used to generate the balance sheet and financial statements using the general ledger. (O'Brien, 2004)

The following illustration describes several aspects of accounting and how an ERP-based system might be useful in enhancing these business processes.

Accounts payable deals with vendors and the transaction with the vendors. Every vendor has to have the master record, which contains important data about the vendor, such as their name, address, and bank information. Accounts payable is also responsible for the payments. Depending on the purchases made from the vendor, the user can make the necessary payment. For instance, one possibility is to use the automatic payment option where the system checks the vendor connection and offers the best and most efficient payment method for the user, such as cash in 14 days with a 5% discount. This enables the company to use its resources effectively and to the best advantage. Payments are automatically posted to the general ledger, and thus can be seen in the financial statements.

To sum up, ERP-based accounting modules automate routine accounting processes and integrate accounting with other business processes using real-time information. Also, they support the decision making process in the company and enable the company to use its financial resources more effectively and efficiently.

Logistics

Logistics is mainly concerned with activities related to the physical flow of materials, semi-finished and finished products, and materials used for production. Logistics span the whole supply chain, from purchase of the raw material to production on one hand, and from production up to sales on the other. Activities such as purchasing, transport, warehousing, organizing, planning and sales are part of the logistics function. In some respects, this extends the functionality of MRP II systems. In logistics, one of the goals is to minimize the costs of all factors that do not contribute directly to the value-added processes. Examples of such factors are reducing warehouse stocks and storage times (Curran & Ladd, 2000).

The benefits that can be achieved by implementing the ERP-logistics modules are significant for a company. It increases the coordination among suppliers and customers and connects them to the logistics process. Communication and cooperation among suppliers, manufacturers, and customers improves, thus reducing costs, optimizing processes, and leading to higher profits. Effective implementation and use of logistics modules can reduce storage times, improve delivery dates, and reduce the negative aspects of unexpected production failures.

Logistics can be classified into three groups: sales and distribution, production planning, and procurement.

Sales and Distribution

The necessity of using a sales and distribution module has increased because the way sales departments do business has changed in recent years. Companies have had to change and become more customer-oriented in order to stay competitive in the market. This has led to the development of systems, which have increased the importance of the sales and distribution systems. The functionalities supported by this module can be summarized as operative sales support, requests for quotations, orders, credit limit checking, pricing, shipping, billing, and contracts agreements. Rather than explaining all these functionalities individually, an illustration might be more helpful. This also can show how these systems support the new customer oriented approach (Buck-Emden, 2000, p. 235).

A typical order starts with the search for a prospective customer. Once contact has been established with a potential customer, which could be done by telephone or written request, a request for a quotation is generated. A quotation is the offer to the customer, which includes prices, delivery times, terms of delivery, etc. Based on the customer's response, the quotation may need to be updated and resent to the customer. Before updating the quotation, necessary changes can be made to meet the demands of the customer. For instance, the system can check whether the item is in stock or not, which will determine possible delivery dates and give information about the availability of the item. Once the customer makes a sales order, the system creates delivery proposals so the user can choose their preferred method of delivery, and then the system informs relevant departments. Finally, the billing function is invoked and the invoice is sent to the customer. Any necessary changes in the system are simultaneously performed, continuously updating functions such as inventory, accounts receivable, and so on.

Even this simplified example shows how important such a system is for sales, and how it can support the whole sales process in an integrated way. Without such systems, it is very difficult to achieve efficiency in sales and distribution because even this simplified example cannot be realized quickly.

Production Planning

As in sales and distribution, the trend in production planning is also towards a more customer-oriented approach. Thus, the production function has to be more flexible and companies have to be ready to easily adapt and change to fulfill the demands of the customer. In the production process, the development times for specific products have shortened drastically. In the mobile phone industry, for example, the development of the next new product starts even before the most recently developed product is introduced to the market. This clearly shows that companies have to be quick to adapt to changes. Another reason is the shortening of life cycle times. In the past, a product could be on the market for several years, whereas now many products survive in the market for only a few months. As a result, the development of new products is of major importance for a company. However, companies can only sustain a high level of competitive advantage if they can continue to achieve high quality and lower production costs.

Clearly, without a suitable production planning module, the flexibility and adaptability needed to attain high product quality and low costs is not realizable. Thus, companies who are manufacturing-oriented must use production planning modules.

Many production planning modules are based on MRP II. What distinguishes these systems from MRP II systems is their integration of production facilities with other components, such as sales, to ensure a customer orientation. Using this system, companies can manufacture their products based on customer needs, which can also be defined as made-to-order production. On the other hand, project-oriented production in special machine manufacturing is also possible, as in this case the system assists in scheduling and coordinating activities, and assures a shorter delivery time.

To sum up, without these systems, it would be very difficult to achieve the desired production flexibility and to overcome the challenges of customer orientation. Some industries may not need production planning systems, but for those that do, these modules are invaluable in helping them achieve a competitive advantage (Curran & Ladd, 2000).

Procurement

The process of buying goods and services to support business operations is known as procurement. In the ERP context, however, procurement is a broader concept that includes purchasing goods, inventory control, warehouse control, supplier selection, and supplier performance control. A company's procurement systems helps to optimize the flow of goods and to coordinate procurement with other business units such as financials and sales.

Procurement systems offer various functionalities. For instance, supplier evaluation and supplier performance management systems allow a company to purchase goods with the lowest price and highest quality. Suppliers make their offers by creating quotations, and the most suitable supplier can be chosen from amongst these. This system also optimizes the procurement process, as by giving purchase authorization to employees, procurement can be done more efficiently and effectively.

To illustrate this, consider the following example:

First, a purchase request must be submitted. According to the requirements, a purchase request will be prepared and a search performed in the already existing contracts with suppliers. If there are no suppliers for the required goods, there will be a request for quotations to find the best supplier for the product. Once a choice has been made, a purchase order will be created and sent to the supplier. In the purchase order, information such as delivery details and payment is given. The supplier will then send a shipping notification. After the items are received, these are recorded in the inventory system and stocks are updated immediately. After receiving the goods and invoice, the system will send information to financials to authorize payment, and to production and sales to inform them that the items are now available for further use.

With the support of the procurement systems, jobs which used to take a long time can now be done quickly. Purchasing is both more efficient and more effective. Companies can easily choose among several suppliers, and thus optimize their purchasing process.

Human Resources

Human Resources (HR) systems, similar to accounting and logistics, aim to utilize the human resources of a company more efficiently and effectively. Human resources have always been thought of as a department that is isolated from other core businesses of the company, but the increasing importance of finding and keeping high quality employees has made HR systems very important for many companies.

A typical HR system includes recruitment, administration, training and development, and compensation functions. Old systems were often limited to administration and compensation (i.e., the company's payroll) and were simply used to create pay checks and payroll reports and for keeping personnel records. However, the new functionalities available in ERP systems have extended this to recruiting, selecting and hiring personnel, employee self service, time management, and many other functionalities. The following scenario provides an example of how an HR system operates:

A typical recruitment scenario begins with the necessity to recruit new workers. To initiate this process, the personnel department might make a request for recruitment. Once the new positions have been advertised, the applications received are evaluated and those who are not eliminated are placed in a candidate pool. The most suitable applicants are then chosen from among the applicants in the pool. This whole process is monitored. After the vacancies have been filled, the process is complete (Curran & Ladd, 2000, pp. 167-174).

Another useful functionality offered by HR systems is the use of "corporate portals." Here, employees can use these portals to retrieve information about themselves, viewing information such as their benefits, checking their monthly payments, verifying how many hours they have worked, and updating their personal data. They can learn about the procedures of the company and also find out who is responsible for a particular HR function. By having access to management information systems, they gain a broader perspective of how the company operates and can thus contribute more. Employees can feel they have more of a stake in their company's success by getting involved in company issues and by having access to information on company policies online (Soluziona, 2005).

HR systems aim to convert HR management from a cost center to a strategic partner. This can help the company to gain a competitive advantage and extend its capacity to attract, develop, and retain key employees. Operational activities are also optimized by HR systems, thereby allowing the company to focus on strategic issues such as planning and personnel development programs.

Other Modules

The above mentioned systems and modules form the core functions of a typical ERP system. However, there are also some cross functional systems that aim to extend the functionalities of ERP systems, namely Supply Chain Management (SCM), Customer Relationship Management (CRM), Supplier Relationship Management (SRM), and Product Lifecycle Management (PLM). These systems help to extend the features of ERP systems (see Figure 3).

Supply Chain Management

Supply Chain Management systems optimize the flow of products in the supply chain. An SCM system supports and manages the supply chain from concept to market. The goal is to control the supply chain and act efficiently. On one end it manages the company's relationships with its customers, while on the other it manages its relationships with its suppliers. Many features of SCM are already offered by ERP systems. Thus, SCM systems extend the functionalities of ERP systems (O'Brien, 2004).

Figure 3. Cross-functional systems and their interaction with each other

Customer Relationship Management

The focus of CRM systems is to foster integration with customers. Companies use CRM systems to attract and retain customers. To successfully implement this strategy, many companies build special databases which contain information about their customers, so that salespeople, managers, or even the customers themselves can access the information. CRM systems aim to manage the company's relationships with customers, and thus its functionalities intersect with the SCM systems. However, CRM systems focus more on customer relations than do SCM systems.

Another important issue that bears mentioning is the difference between operational and analytical CRM. Operational CRM deals with business processes such as customer contacts, including sales, marketing, and purchases. The necessary information to deliver these tasks is gathered by using operational CRM and the back-end connection. For instance, when a customer calls, relevant data about the customer can be retrieved easily and better service can be provided (Wikipedia, 2005).

On the other hand, in analytical CRM, data gathered about customers are analyzed and used for market segmentation, to define customer needs and requirements, or to develop sales strategies. For instance, by analyzing the buying habits of a specific customer group, targeted selling strategies can be implemented.

Supplier Relationship Management

Supplier Relationship Management is a comprehensive approach to managing an enterprise's interactions with the organizations that supply the goods and services it uses. The goal of supplier relationship management is to streamline and make the processes between an enterprise and its suppliers more effective, just as customer relationship management is intended to streamline and make the processes between an enterprise and its customers more effective (Spencer & Kumar, 2005).

Product Lifecycle Management (PLM)

PLM systems support the entire lifecycle of a product from concept to market, and may even include recycling the returned products. They tend to be found primarily in manufacturing companies. PLM software often uses computer-aided design tools to support and manage product development. PLM features are also integrated in ERP systems in production planning systems. Similar to

SCM, PLM extends the use of these systems and focuses on the product. PLM systems require a high degree of integration internally and externally because the product development process is often shared between the manufacturer and a large number of suppliers. Internally, the PLM system often delivers basic data as well as documents and supports a wide range of processes affecting different internal and external departments.

IT Architecture of an ERP System

Now that we have discussed the features of an ERP system, it is time to understand the underlying technology involved in an ERP system. We will first define some terms and then provide a few real-world examples to show how ERP systems are used.

Mainframe Architecture

Before the widespread use of personal computers, many systems were built on mainframes which were very large and expensive computers that supported thousands of users. There was a central, huge computer which had an amazing processing capacity. The terminals used to communicate with the mainframes could only display what the computer sent, and send back to the computer what the user entered. It was easy to perform maintenance on such computers, but any failure of the main computer could paralyze the whole company. With the emergence of graphical user interfaces (GUI) such as Macintosh and Windows, the expectations of users increased. To perform the enterprise applications that such GUIs demanded, a shift towards client/server architecture was needed (Langenwalter, 2000).

Client/Server Architecture

Client/server architecture involves dividing the application software and the processing load between two or more computers. Let us examine what happens when a client requests a service from the server. First, the desktop PC which is responsible for the user interaction activities requests a service from the application server, which could be a request for information about a customer. The application server does not have the data available, so it assumes the role of a client as it requests the data from the database server. Whether a machine acts as a client or a server thus depends on the kind of transaction the machine

Figure 4. Two-tier client/server architecture (Sadoski & Comella-Dorda, 2000)

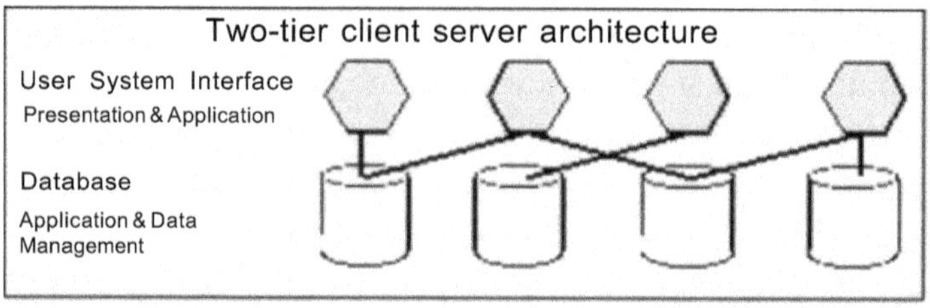

is involved in. The goal of the client/server architecture is to improve the usability, flexibility, interoperability, and scalability of applications. With the aid of this technology, a new user can easily learn to use the system. The system can be adapted to fit into new environments more easily and the ability to exchange information among several systems is enhanced. These features make a client/server architecture superior to a mainframe architecture.

The client/server architecture started with a two-tier architecture, where the client directly connects to the server and exchanges information. The processing is split between the client and the server. A typical use of this architecture is the Web browser on a PC and a Web server available on the Internet (such as www.yahoo.com). The same architecture is used in an e-mail system, with the e-mail client being a tier and the e-mail server being the second tier in the architecture.

As shown in Figure 4, application and data management functions can both reside on the server, while the client computer provides the presentation and application functions. Let us look at how this works for an e-mail application such as Outlook Express e-mail. Here, the presentation and application functions reside on your computer. You set up options and accounts functions so that this application queries an e-mail server at specified intervals and brings any e-mails residing on the server to your computer. You may be able to access the e-mail from the main e-mail server by logging on to it directly (mainframe architecture) or using Outlook Express from your computer to retrieve e-mails (two-tier client/server architecture).

A three-tier client/server architecture is normally used for ERP systems. These typically consist of presentation (user interaction), application, and database. By adding an extra layer, the direct connection between the client and database server no longer exists, as communication occurs solely between the application servers and database servers. This extra tier helps to increase the flexibility, interoperability, and scalability of the client/server architecture. A typical three-

Figure 5. Three-tier client/server architecture (Sadoski & Comella-Dorda, 2000)

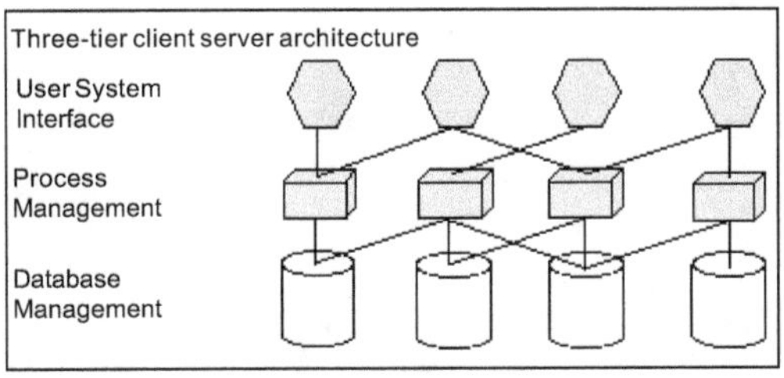

tier architecture is shown in Figure 5. An example of the use of this architecture is with airline reservation systems. You use a product such as Expedia (user system interface) to submit a query about possible tickets to serve your travel needs. The user interface system sends the queries to a process management system that routes the queries to different databases. The query about the reservations available might come from the databases of multiple airline carriers, such as Delta, Lufthansa, US Airways, and so forth.

The middle tier in Figure 5 may be split into many units, effectively making it a multi-tier architecture. It is now possible to support hundreds of users working simultaneously by implementing three-tier and multi-tier architectures, which provide the underlying technology supporting an ERP system.

Until the implementation of three-tier client/server architecture, companies were unable to use ERP systems to their full potential, as ERP relies on this architecture. Figure 6 illustrates the effect of recent developments in technology. Mainframe architecture was used until the 1980s, but as the complexity and costs increased, a new trend became popular. This was the client/server technology. The implementation of client/server technology occurred with the implementation of three-tier architectures in the early 1990s. This opened the way for systems such as ERP, SCM, and CRM, which entered the market and were implemented by many companies. However, complexity and cost are again on the rise and a new trend in the market is becoming evident, namely Service-Oriented Architecture. This is likely to be the wave of the future for commercial computing applications.

Service Oriented Architecture (SOA)

Service Oriented Architecture (SOA) is an information technology that defines the services designed to support business functions that are available in a

Figure 6. Wave model showing the trends for several technologies (Agassi, 2004)

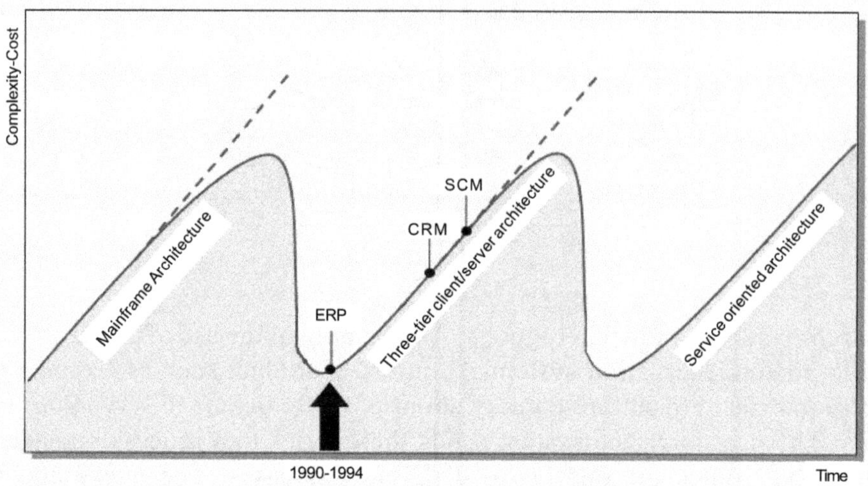

network, such as the World Wide Web. A service is a business function, such as a credit card authorization. In this technology, there are service providers who provide specific services. The client or the service consumer seeks specific services that are offered by service providers and then uses these services (Mahmoud, 2005).

If a service consumer needs a specific service, such as credit card authorizations, he/she first seeks out available providers. After identifying a list of service providers, they negotiate a contract for the service to be provided and agree on a fee. The service consumer can then use this service as needed.

SOA architectures that are based on the World Wide Web are referred to as Web services. To maintain interoperability, specific standards are used on the Web, such as XML technology. This makes SOA possible on the Internet. The logic is the same as above. As soon as a Web service is found, the client can make requests and the service provider will give a response to this request. This could be a request for a credit card check, currency conversion or purchase order process. What makes this type of Web service superior to previous technologies is that it increases flexibility, interoperability, and scalability. Finally, an application need not be a stand-alone solid software system, but a well-structured configuration of services. In such cases, the services can be provided as a client/service architecture, although a legacy application on a mainframe may act as the background server.

Figure 7. SOA interaction between service providers and consumers (Mahmoud, 2005)

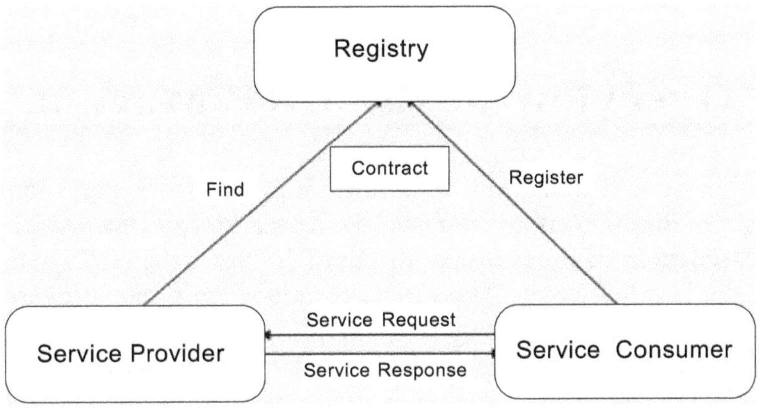

Work Flow

This term is often used when designing ERP systems and is worth mentioning here. Work flow defines who is responsible for what. To understand how this functions, we can look at the example of an employee who is authorized to purchase material up to $1,000 without waiting for approval. Of course, this limit can be changed, but for now let's assume it is $1,000. For purchases between $1,000 and $10,000, the employee cannot make a purchase without prior approval by a supervisor. If the amount exceeds $10,000, the approval of the Vice President might be necessary. This is known as work flow. All these rules are implemented in the work flow system and the system controlling the work flow will apply the rules. To define who will be responsible for a purchase, a workflow system might obtain personnel data from the HR system where the organizational structure is modeled. The workflow system integrates the different functionalities of an ERP system.

Often the work flow system includes a document management system; this does not simply eliminate the need for paper based documents being sent from one department to another, but also saves a lot of time. If used properly, it will help to optimize and monitor the processes. It will also help to define and implement the policies and procedures of the company. Of course, to be able to implement a work flow system, a company must already have well-defined procedures and it has to be clear who is responsible up to what level. Without clear definitions and delineations of authority, it is impossible to implement a work flow system (Langenwalter, 2000).

Overview of the ERP Market

In this section we will provide an overview of the ERP market and compare the major ERP vendors. This is a very fast growing market, and many ERP vendors have increased their revenues from millions of dollars in the 1990s to billions of dollars in the last few years. The global revenues from the software used to automate financial, human resources, procurement, and other business operations was $26.7 billion in 2004 according to International Data Group, up from $25 billion in 2003, and shows the steady increase of ERP revenues worldwide. In addition, the top 10 ERP vendors collectively increased their market share by 2% each year, beginning from 2001. It is estimated that approximately 50% of the market share belongs to the top 10 vendors. Due to a series of acquisitions and collaborations in the last few years, the top five ERP vendors' ranks have changed somewhat, but SAP remains the largest ERP vendor, with over $10 billion in annual revenue. The other important ERP vendors are Oracle, PeopleSoft, J.D. Edwards, and Microsoft. The focus in this section will to discuss these companies in turn, with a detailed section on the largest ERP vendor, SAP (Staff Writers, 2004).

During the 1990s, the ERP vendors focused on big companies and implemented ERP systems tailored for their specifications. Most of the Fortune 500 companies implemented ERP systems in the 1990s. However, since 2000 the structure of the market has evolved and the emphasis on mid-market and SMB (small- and medium-sized businesses) has increased. As a result, specific solutions targeting these groups have gained in importance. The definition of SMBs or mid-market businesses varies among suppliers, but generally SMBs are defined as businesses with less than $200 million in revenues, while mid-market businesses are defined as those with annual revenues between $200 million and $1.5 billion. Large industries are those that generate more than $1.5 billion in revenue each year.

ERP vendors provide industry specific solutions for large enterprises. For instance, the requirements of the automotive industry differ considerably from the requirements of the chemical industry, so ERP vendors must provide industry specific solutions to meet each of these sets of requirements. In Table 1 we list some of the industries supported by major ERP vendors (SAP AG, 2005a).

Another important market worth mentioning is the software that provides additional functionality beyond that provided by ERP systems. Some of these

Table 1. Industry specific solutions provided by major ERP vendors

Aerospace & Defense	Higher Education & Research	Oil & Gas
Automotive	Industrial Machinery & Components	Pharmaceuticals
Financial Services	Insurance	Postal Services
Chemicals	Life Sciences	Professional Services
Consumer Products	Logistics Service Providers	Public Sector
Defense & Security	Media	Railways
Engineering, Construction & Operations	Mill Products	Retail
Healthcare	Mining	Telecommunications
High Tech		Utilities
		Wholesale Distribution

software products were discussed in an earlier section, namely CRM, SCM, SRM, and PLM. Another new product is called Business Intelligence software. This analyzes data and compares the results for various "what if" scenarios or predictions. There are also several tools that may be used to provide additional functionality, thus extending the scope of ERP systems. The list of major vendors for these products is listed in Table 2, classified according to the Gartner Research Magic Quadrants system. The choice of ERP vendor varies from country to country or from industry to industry, but there is a general trend when choosing an ERP vendor. One of the most important criteria is the maturity and the stability of the ERP vendor. In particular, big companies choose big ERP vendors in order to enhance their guarantee of success. Also, the viability of the ERP vendor, in terms of its financial strength and market share, is a very important criterion when choosing an ERP vendor. The technology used to implement the product is also vital.

Cisco successfully implemented an ERP system in a very short time. When they were looking for an appropriate ERP vendor, their strategy was to build as much knowledge about the technology as possible by leveraging the experience of others. They also used benchmarking results of important research sources to focus on five vendors. One week later, they narrowed this down further to two big ERP vendors, one of which was Oracle. The reason they chose these two big ERP vendors was stated as follows: "We decided that we should not put Cisco's future in the hands of a company that was significantly smaller than we were." This clearly shows a big company's attitude when choosing an ERP vendor.

Table 2. Proprietary software solution providers

Supply Chain Management (SCM)	Product Lifecycle Management (PLM)	Customer Relationship Management (CRM)	Business Intelligence (BI)
i2 Technologies	Agile Software	Epiphany	Actuate
Logility	Centric Software	Pegasystems	Business Objects
Manugistics	Dassault Systems	Siebel	Cognos
SAP AG	Formation Systems		Hyperion Solutions
	Invensys/Baan		Informatica
	Matrix One		Information Builders
	Oracle		MicroStrategy
	PTC		SAS
	SAP AG		
	UGS PLM		

Cisco decided to work with Oracle after extensive discussions with the two vendors. The rationale behind this decision was (a) the company's manufacturing emphasis and the offer of the ERP vendor matched; (b) the long term promises received from Oracle were satisfactory, which shows that support is an important issue for a company; and (c) Oracle is located close by, thus offering more flexibility to Cisco. The reasons were thus not necessarily either economic or technical (Cotteleer, Nolan, & Austin, 1999, pp. 4-5).

You will find many smaller vendors with a large market share in different countries, but their market focus is generally on SMBs.

SAP

SAP is the leading ERP vendor in the market, with an annual revenue of more than $10 billion. SAP was founded in 1972 under the name "Systemanalyse und Programmentwicklung" (Systems Analysis and Program Development) in Mannheim, Germany by five former IBM employees. Their vision was to develop and market standard enterprise software that would integrate a wide range of business processes and provide real-time interaction (Langley, 2003).

SAP software is implemented in almost every industry mentioned in Table 1. SAP's strength lies primarily in its ERP solutions, but in the last few years the company has extended its product line by offering components such as CRM and PLM. Under the name mySAP, the company markets products in almost every business area possible.

SAP is well established in the large enterprise market, where its solutions are deployed by many companies. In the last few years, however, SAP has realized the importance of the mid-market and SMBs, and has started offering solutions tailored for this market segment. By the year 2004, SAP solutions had been deployed by more than 26,000 user organizations, in over 120 countries, with more than 88,700 installations (AEIOO, 2005).

SAP has always been seen as a very strong company with high quality products. It has been in the market since the very beginning and has grown steadily, enjoying a high level of cooperation with its customers. Its maturity and stability are its greatest strengths. Unfortunately, for many years it has been seen as a company that only large enterprises could afford, although recent figures show that 57 % of its customers are businesses with less than $500 million in revenue (Mc Dermott, 2004). With its efforts to sell in the mid-market and SMB market, SAP will likely continue to be a dominant ERP vendor.

Strength and Weaknesses

SAP has focused on producing products and services in the business application market since 1972, and it has long been a leader in this market. Its early business solutions have now evolved into Web-enabled ERP solutions. Its long history in the market, in addition to the successful strategies it has implemented, have resulted in SAP's domination of the market. Even after mergers involving some of its competitors, SAP remains the dominant ERP vendor worldwide, with approximately 55% of the market share. The continuous improvement of its product portfolio is a key success factors for SAP. Its latest development is the Web-enablement of its products and service-oriented architectures. SAP has a strong integration potential, which enables third party vendors to integrate their software with the solutions of SAP.

The company's main weakness is its negative image, where SAP is perceived as being a company that works mainly with large enterprises. Generally, SAP solutions are seen as difficult and very costly to implement, and total cost of ownership is seen to be relatively high in terms of the initial product and implementation costs (AEIOO, 2005).

Oracle, PeopleSoft, and J.D. Edwards

Oracle Corporation is an enterprise software company that focuses on database and infrastructure software, including application servers, collaborative software, and development. The company also offers business applications solutions, known as the Oracle E-Business Suite. This suite offers solutions for specific industries as well as small- and medium-sized businesses. Their diverse product portfolio enables the company to serve many different customers.

PeopleSoft and J.D. Edwards are included under this heading because of several mergers that occurred in 2004. Oracle had been trying to buy PeopleSoft since 2003, but because of legal issues and antitrust laws the process took some time. In the meantime, PeopleSoft acquired J.D. Edwards. Oracle was finally able to take over PeopleSoft in 2005. As a result of this merger activity, there are now three products offered under the Oracle brand name.

PeopleSoft Solutions offers fewer industry-specific solutions compared to the Oracle E-Business Suite. PeopleSoft Enterprise offers industry-specific solutions for healthcare, education, and life sciences, and concentrates on small- and medium-sized businesses.

J.D. Edwards EnterpriseOne is an industry-specific solution that mainly focuses on manufacturing, construction, distribution, service, or managing products. With low cost of ownership and specific solutions, it is ideal for SMBs.

The challenge for Oracle is to support multiple product lines while simultaneously devoting a great deal of effort to merging these different ERP products. Oracle must ensure that the support for the existing customers of PeopleSoft and J.D. Edwards will continue in order to prevent uncertainty and protect its customer base. Meanwhile, major competitors of Oracle such as SAP and Microsoft have started a campaign to lure away the customers of the acquired companies PeopleSoft and J.D. Edwards. SAP, for example, has introduced the Safe Passage program, which is specifically designed for the customers of PeopleSoft and J.D. Edwards who might be worried about the continuing support for their products. SAP was able to sign up 21 customers for the Safe Passage package of maintenance and migration support in the first half of 2005 alone (Sayer, 2005).

In response, Oracle has published a timeline explaining how it is planning to continue to support existing customers. For instance, PeopleSoft products will be supported through 2013. Also, Oracle wants to use the synergy effect of this merger to provide a superior product portfolio by combining the strengths of the combined products. A roadmap has been already published showing the combination efforts under the name "Fusion." In the "Fusion" standard, Oracle aims to incorporate the best features and usability characteristics from Oracle and PeopleSoft products (Oracle Co., 2005).

Strengths and Weaknesses

Oracle is financially a strong company, which gives it the capability to absorb these mergers and support both their existing and newly acquired customers. Oracle has dominated the database management market for years, although this may also be seen as a weakness because historically Oracle has been positioned as a relational database management system vendor.

The new acquisitions have extended the product portfolio of Oracle and have also given it the option to compete much more strongly in the field of proprietary software vendors, with products such as CRM and PLM. Unfortunately, these acquisitions have also placed a considerable burden on Oracle's financial strength, and the necessity to support the acquired customers might be costly for Oracle.

Microsoft

Microsoft entered the business solutions market relatively late compared to its competitors. It was slow to realize the potential of the ERP market, but has since invested heavily to gain a significant market share in this huge market. Microsoft entered the market several years ago by purchasing Great Plains Software and Navision, which afterwards were combined to form Microsoft Business Solutions Suite. In contrast to SAP and Oracle, Microsoft has mainly focused on serving SMBs.

Strengths and Weaknesses

Microsoft is a software giant, dominating the market in desktop applications and operating systems. Trusting in this power, Microsoft did not hesitate to enter the ERP market by spending billions of dollars to reach its competitors' level. However, lacking the historical experience and market share of its competitors, Microsoft has to rely on this strength to be competitive in this environment. Microsoft has not focused on large enterprises, and thus does not offer as many industry specific solutions as its competitors. Microsoft's Navision is more focused on small- and medium-sized businesses, with a general strategy of concentrating on the mid-market.

Microsoft noticed that many of its competitors lacked a good graphical user interface. For many users, it was thus quite difficult and time consuming to learn to navigate the software packages of many ERP vendors. Microsoft exploited this weakness and built Office-like interfaces that new users could easily learn and navigate. Figure 8 shows an example of the graphical user interface in

Figure 8. Microsoft Navision Graphical User Interface (Microsoft, 2005)

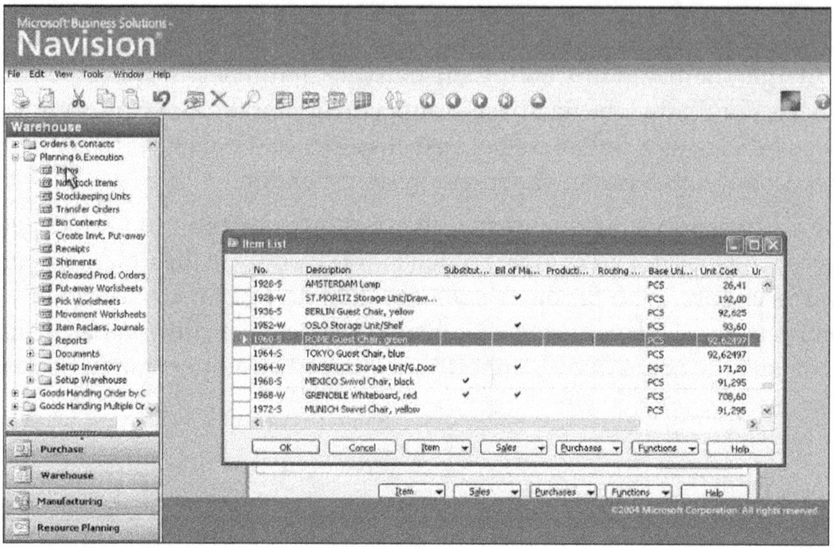

Microsoft Navision. Users of Microsoft's e-mail software, Outlook, will easily recognize the similarity.

Another improvement Microsoft introduced was the potential for integration of the business applications with the desktop applications. For instance, meetings that have been arranged with customers can be synchronized directly into Microsoft Outlook in their ERP system, which improves the connection with the customers. Revenue reports or payment offers to the vendor can easily be exported to Excel or PowerPoint, thus providing the user with more functionality. Editing is straightforward and data can easily be prepared for a presentation. Microsoft also realized that the data stored in desktop applications could not be entered into the system automatically and generally had to be reentered. Microsoft created solutions to overcome these difficulties by enhancing the capability for data exchange between desktop applications and business applications.

The company's competitors have become aware of this threat and several of them are now cooperating with Microsoft. The closest cooperation exists with SAP. A few years ago there were rumors of a possible merger, but both companies decided that it was not in their best interests to merge and that they should continue to work together as strategic partners. The two companies have a 10 year agreement to integrate Microsoft's .Net platform with SAP's NetWeaver platform, which both aim to integrate their users with other applications. This enables Microsoft Office users to easily connect to SAP applications. This partnership will also define Web service standards in the future (Fox, 2004).

Details about SAP R/3: The Leader in the Market

The case studies used in this book are focused on SAP implementations, so it is important to understand the fundamentals of SAP systems and how they are implemented. This section focuses on the SAP R/3 system and explains its most important features.

History

SAP's business application solution is the SAP R/3 system, where "R/3" stands for real-time system, version 3. The R/3 system evolved from its predecessor the R/2 system, which was designed for use with mainframes. The R/3 system is built on a multilayer client/server architecture and is thus very different from its predecessors.

Originally, in 1987 IBM introduced the SAA (system application architecture) which aimed to create platform-independent applications with a graphical user interface. Unfortunately the lack of connectivity with existing systems hindered the success of SAA and the concept failed. However, SAP was able to use this revolutionary concept to develop the R/3 system, which was able to use open systems. In 1992, SAP introduced the R/3 system, which was one of the few business applications that used a graphical user interface. The system was constantly updated with improved versions and has evolved into a very effective Web-oriented and Web-enabled business application system (Buck-Emden, 2000).

Modules

The R/3 system offers solutions in three main modules: accounting, logistics, and human resources. Table 3 shows these three main categories and their subcategories.

Table 3. Modules of the SAP R/3 system (Buck-Emden, 2000, pp. 230-240)

ACCOUNTING	LOGISTICS	HUMAN RESOURCES
Financial Accounting	Sales & Distribution	Personnel Management
Investment Management	Production Planning	Organizational Management
Control	Materials Management	Payroll Accounting
Treasury	Plant Maintenance	Time Management
Enterprise Control	Quality Management	Personnel Development
	Service Management	

Accounting

The accounting module is of vital importance and is therefore the module that is implemented first in a typical SAP implementation. This module consists of five specialized areas: financial accounting, investment management, control, treasury, and enterprise control. Accounting deals with every interaction that a company has with its environment concerning financial issues, as well as with the internal processes involved in consumption and production. Areas such as accounts receivable and accounts payable (under financial accounting control) cover interactions with vendors and customers. Control focuses on controlling the company's internal processes. Investment planning, budgeting, and control ensure that a company has a better handle on their investments. The accounting module also controls and manages the legal issues connected with financial transactions.

Logistics

Logistics is the largest module in the R/3 system. It consists of six specialized areas: sales and distribution, production planning, materials management, plant maintenance, quality management, and service management. Logistics allows real-time monitoring of raw materials, finished goods, work-in-progress, and so forth. Using logistics functionalities, it is possible for the user to enhance just-in-time production and point-of-sale technology. It also helps to optimize warehouse management and inventory control functions. The area of sales and distribution is dedicated to customer side transactions, focusing on subjects such as requests for quotations, pricing, and shipping. Materials management deals with interactions with vendors by optimizing requirements planning, inventory management and warehouse management, or issues such as purchasing and invoicing. Production planning controls processes concerned with production using functions such as material requirements planning or project management.

Human Resources

The human resources area consists of five specialized areas: personnel management, organizational management, payroll accounting, time management, and personnel development. Realizing the importance of the HR module in designing optimal business processes, R/3 systems' functions for human resources are tightly linked to the components of financial accounting, control, and production planning. International features such as currency conversion and variable taxing enables multinational companies to manage widely dispersed HR functions from a central location effectively (Center for Virtual Organization and Commerce, 1998).

Best Practices

ERP systems are based on best practices defined by the vendor, so the company must adapt its business processes to fit the processes provided by the vendor. SAP systems has managed to create processes that are based on best practices by incorporating customization options that allow their customers to tailor their specifications up to a certain limit. Within the R/3 system, over 800 best practices are included that could be examined and implemented by a customer in the implementation and customization phase. For this purpose, a Business Process Reengineering team defines the best practice for the company and customizes it in the system. The user then has the option to choose from among many available best practices. For instance, the user can examine a made-to-order process scenario and identify several steps that are irrelevant for its own processes. The irrelevant steps can easily be deactivated in the system, thus optimizing it for the user. As a result, processes designed for the customer can be realized easily and can be implemented with minimal effort.

Enabling Technologies

ALE

The technologies introduced by SAP have improved the R/3 system and allow it to be used in almost every possible environment. One of the technologies introduced by SAP is Application Link Enabling (ALE), which improves the integration potential of the R/3 system. The purpose of this technology is to integrate an R/3 system with non-R/3 systems, as well as with other R/3 systems.

In this technology, application systems are loosely coupled. This simply means that the data exchange does not happen synchronously. When data is entered in one part of the system, other parts of the system may not be updated immediately. The ALE scenario defines the exchange process and ensures that the update does eventually happen. This means that data can be exchanged asynchronously. Using this technology, company-wide applications such as accounting and human resources can be managed at the company's headquarters, while production and logistics can be managed in decentralized plants.

Web-Enabled Technology

In the future, technology for enterprise applications will be based on a service-oriented architecture or Web-enabled technology. In order to accommodate this approach, SAP has introduced several new technologies such as SAP Netweaver and a new concept called Enterprise Application Integration.

SAP Netweaver provides the integration and application platform that serves as the foundation of SAP's Web-enabled solution called as mySAP Business Suite. Solutions such as Enterprise Portals or Knowledge Management enable and enhance the integration potential within the company. It also enables easy integration with other systems, such as .Net from Microsoft or WebSphere from IBM. Netweaver is the underlying technology for service-oriented architecture. SAP xApps and mySAP Business Suite are both compatible with this technology and will form the basis for SAP's business application solutions for the future.

R/2 was the technology of the mainframe era; R/3 was designed to serve the client/server architecture era. In the future, Web-enabled, service-oriented applications will be based on SAP xApps technology and mySAP Business Suite. Chapter X discusses this prospect in more detail.

Chapter Summary

This chapter discussed the historical basis of ERP systems and traced their evolution. It then listed the components of an ERP system, namely financials, logistics, and human resources, and described these components. The chapter went on to describe today's client/server architecture and how it works with ERP, along with the service-oriented architecture that will form the basis of future ERP applications. The market for ERP systems was then described and the major vendors and their products explained. The major modules of SAP's R/3 were outlined. This chapter has provided you with a strong introduction to ERP systems to give you a good overview of their systems, functions, and modules.

Student Assignments/Questions

1. Explain the evolution from MRP to ERP.
2. Give a brief overview of the components of an ERP system.
3. Explain the main characteristics of a client-server-architecture.
4. Give possible reasons for choosing a step-by-step implementation.
5. Do you think that it is correct to say that there are very few vendors in the ERP market?
6. In what situation might a company need to use the ALE technology developed by SAP?

References

Army Enterprise Integration Oversight Office (AEIOO). (2005). *Vendor profiles*. Retrieved July 22, 2005, from http://www.army.mil/aeioo/erp/vendors1.htm

Agassi, S. (2004). *Emergence of the smart enterprise suite*. Paper presented at SAP Developer Network, Elearning. Retrieved May 7, 2005, from https://www.sdn.sap.com/sdn/elearning.sdn?class=/asug/portal_business/eClass/504.htm

Austin, R. D., Cotteleer, M. J., & Escalle, C. X. (2003). Enterprise resource planning. *Harvard Business School*, 1-8.

Buck-Emden, R. (2000). *The SAP R/3 system: An introduction to ERP and business software technology*. London: Addison Wesley.

Center for Virtual Organization and Commerce. (1998). *Best practices*. Retrieved June 29, 2005, from http://isds.bus.lsu.edu/cvoc/learn/bpr/mprojects/bp/bpsap.html

Cotteleer, M. J, Nolan, R. L., & Austin, R. D. (1999). *Cisco Systems, Inc.: Implementing ERP. HBS case study (pp. 1-18).* Harvard Business School.

Curran, T. A., & Ladd, A. (2000). *SAP R/3 business blueprint: Understanding enterprise supply chain management* (2nd ed.*)*.Upper Saddle River, NJ: Prentice Hall PTR.

Fox, P. (2004). Big isn't better. *Computerworld, 38*(25*)*, 20.

The Free Dictionary. (2005). Manufacturer resource planning. Retrieved June 22, 2005, from http://computing-dictionary.thefreedictionary.com/Manufacturer%20Resource%20Planning

Langenwalter, G. A. (2000). *Enterprise resource planning and beyond: Integrating your entire organization*. New York: St. Lucie Press.

Langley, N. (2003, April 22). Demand for SAP skills on the rise: Industry overview. *Computer Weekly,* 60.

Mahmoud, Q. H. (2005). Service-oriented architecture (SOA) and Web services: The road to enterprise application integration (EAI). *Sun Developer Network*. Retrieved July 12, 2005, from http://java.sun.com/developer/technicalArticles/WebServices/soa/

McDermott, B. (2004, June 14). CRN interview-Bill McDermott. SAP America, overview. *Computer Reseller News.*

Microsoft. (2005). *Microsoft navision demos*. Retrieved August 2, 2005, from http://www.microsoft.com/BusinessSolutions/Navision/demos.mspx

Oracle Co. (2005). Oracle applications: Committed to your success. *An Oracle White Paper*. Retrieved July 23, 2005, from www.oracle.com/applications/suites.html

O'Brien, J. A. (2004). *Management information systems: Managing information technology in the business enterprise* (6th ed.). New York: McGraw-Hill.

Sadoski, D., & Comella-Dorda, S. (2000). *Three tier software architectures*. Retrieved July 4, 2005, from http://www.sei.cmu.edu/str/descriptions/threetier.html

SAP AG. (2005a). Industry specific solutions. Retrieved July 28, 2005, from http://www.sap.com/industries/index.epx

Sayer, P. (2005). SAP revenue and income rise on software sales. *Computer World*. Retrieved July 21, 2005, from http://www.itworld.com/Tech/4535/050721saprevenue/

Soluziona. (2005). *ERP technology: Finance, logistics & HR*. Retrieved July 2, 2005, from www.soluziona.com/htdocs_en/areas/consultoria/servicios/finanzas_logistica/area_rrhh.shtml

Spencer, D. P., & Kumar, S. (2005). *Order management. InfoSys Technologies Ltd*. Retrieved April 12, 2005, from http://whitepapers.tmcnet.com/detail/RES/1115745398_194.html

Staff Writers. (2004). ERP market benefiting from the IT spending increase. *TechWeb News*. Retrieved May 18, 2004, from http://www.itnews.com.au

Tuteja, A. (2005). *Enterprise resource planning: What's there in it!* Retrieved June 21, 2005, from http://www.geocities.com/CollegePark/Library/6045/erp.html

Wikipedia. (2005). *Customer relationship management*. Retrieved August 3, 2005, from http://en.wikipedia.org/wiki/Customer_Relationship_Management#Operational_CRM

Chapter IV

Management Issues in Implementing ERP Systems

Learning Outcomes

- Understand the need for change and the role of ERP systems in change management
- Identify phases in the change management life cycle
- Obtain a broad understanding of the issues involved in implementing ERP systems in each phase of the change management life cycle
- Identify the phases in implementing an ERP system

Overview

Change is the only constant in business and personal lives. Weather patterns may change rapidly during a day and people must adapt to the changes quickly in order to use the weather to their advantage. Many people, especially in underdeveloped countries, who are not informed of impending changes in the weather have to face these change quickly with no time to react. This can lead to major devastation and many deaths, as reported in the cyclone that struck the state of Orissa in India during 1999, leaving about 10,000 people dead and over 10 million homeless. People in developed countries such as the U.S. generally learn about impending disasters early enough (such as the four hurricanes that hit the southern states during 2004: Hurricane Charley, a category 4 hurricane that killed 34; followed by Hurricane Frances, killed 38, Hurricane Ivan, which killed 52, and Hurricane Jeanne, which killed 28) to leave the most-impacted areas for a period of time and return after the disaster has passed. As a result, even though the damage to property may be severe in both cases, the number of people who die is likely to be higher (about 10,000 in India) when people are unaware of the impeding change and smaller (about 130 in the U.S.) when a major change has been forecast and people can take appropriate action to cope with it. Complex information technologies such as the National Hurricane Center systems aid the public in making informed decisions. Radio and T.V. stations play a strong role in keeping the people informed. In contrast, the flow of information was sparse and uneven and no escape routes were available in Orissa, leading to people staying to face the brunt of the cyclone ("Major Storms", 2005).

The recent disaster when Hurricane Katrina hit the U.S. showed how availability of effective information systems alone are not sufficient to avert a crisis. The country was caught unprepared when a whole city such as New Orleans was under water when Hurricane Katrina hit during August 2005 and a levee broke; relief operations did not arrive in time and more than 10,000 people are expected to have died (Coy & Foust, 2005). This example shows that effective planning, leadership, information systems, and implementation are critical in averting a major crisis and reduce human suffering. That is why it is critical that you are aware of the stages of the change management process and get trained to handle the complex situations that you will face in your career, personal life, and in life.

Similarly, change is common in business and must be dealt with on a routine basis. As companies become larger in size and scope, they rely much more on information systems for their operations. In this chapter, we will provide you with concepts that will equip you to assess your company's industrial position, identify its core competencies, and manage change more effectively. The change management life cycle is described and the issues involved in implementing ERP systems are described. Many information technology projects fail because the people who are assigned to work with the new system are simply not willing to accept their new role. The chapter concludes by listing the management challenges in implementing an ERP system.

Change Forces

Most companies operate in a constantly changing business environment. A model developed by Porter (1986) conceptualizes the forces that impact a

Figure 1. Porter's five-forces model (Recklies, 2001)

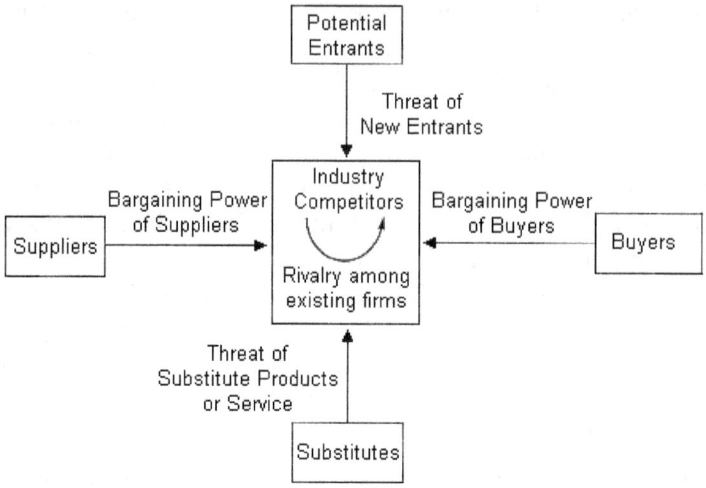

business (Figure 1). In addition to the forces listed in Porter's five forces model, companies often face industry changes due to government or regulatory agencies issuing new laws and regulations.

If present trends continue, the world economy is expected to grow by 50% or more over the next two decades, with international trade a key segment. The dynamic Asian economies are likely to continue expanding, with China emerging as the world's largest economy (Parker, 1996, p. 7). In Europe, the completion of a single market, progress toward economic and monetary union, and institution of a single currency are transforming European businesses. The gradual expansion of NAFTA agreements to Central and South America is a milestone for the United States. Security of systems and people has become a critical global issue due to the terrorist attacks in many countries.

The forces shown in Figure 1 are acting on companies all around the world much more forcefully and quickly than in the past. Massive retail chain stores such as WalMart are gaining a foothold in many countries, thereby introducing consumers to a variety of products at inexpensive prices (Parker, 1996, pp. 192-193). The changes are leading to a redesign of organizational structures using new models: As enterprises move from hierarchical organizations (with a focus on management control) to horizontal organizations (built around teamwork and empowerment), this changes how and where work is done, and who does it. The move towards strategic alliances and virtual enterprises increases the importance of strengthening organizational core competencies, innovation, flexibility, and agility. This compromises any previous assessment that assumed a hierarchical model or did not consider temporary alliances.

Moving from an industrial to an information-based national economy changes the nature of jobs, products, services, and organizational form. This determines if an enterprise is a sustainable business with favorable long-term global opportunities. Moving from a producer focus to a customer focus shifts the intent of service and quality programs, changes the direction of research and development, and creates the potential for distribution channel restructuring and market redefinition. This shift challenges previous planning by questioning whether an enterprise has the core competencies to develop, or the organizational structure to deliver, a product that the customer actually wants. In addition, this brings into question whether the enterprise knows its customers and competitors at the local and global levels.

The competition becomes more intense as IT is used as a competitive advantage by some organizations, thus threatening the status-quo in other organizations. For example, WalMart has used supply chain and ERP systems to sell quality products at an inexpensive price, thereby threatening many other companies such as K-Mart, Caldor, etc. Many of these competitors have either disappeared or merged with other companies in order to compete more effectively with WalMart.

The recognition that IT and computer networks are enablers of social and organizational transformation changes the IT charter from one of aligning with national and business strategies to the need to become an integral part of these strategies. To remain competitive, the enterprise must transform itself and the way it does business by redefining its market, products, suppliers, labor force, business processes, and organizational structure, with each redefinition signaling a corresponding transformation in the use of IT. Previously, businesses seldom even included the IT group in their strategic planning sessions. Instead, they simply advised the IT organization on a need to know basis and only after completing the strategic plan. Today, however, a useful litmus test for detecting whether an enterprise is transforming — as compared to merely making incremental improvements — is if IT is a true business partner in the company's strategic planning process.

Governments are constantly enacting legislation that influences how companies perform and the trend is to liberalize trade across borders, thereby fostering intense competition among companies, people, and products. Multinational companies are being encouraged to identify best business practices and implement them world-wide.

These pressure are forcing companies to change faster than ever. Companies have always had to define and monitor changes in their markets to anticipate threats. Nevertheless, market changes occurred slowly and companies tended to be more concerned about the shifting forces within an industry, allowing them to make fairly accurate short to midterm planning decisions. In today's fast-paced business environment, companies face frequent rapid changes in both their marketplace and in their industry, making it increasingly difficult to predict their future.

Need for Integration and the Role of IT

The forces identified above have made the necessity for integrating the whole enterprise critical. This involves integrating the people, processes, and products of the company so that engineering design, materials and manufacturing functions, and support services work together enabling customer requirements to be met efficiently and inexpensively (Langenwalter, 2000). In order to cope with this difficult environment, companies are seeking the help of information technologies to the point where today's organization can no longer remain competitive without IT. To do this effectively, corporations need to clearly define the position of IT and its impact on the company's operations. It is important at this point to realize, that IT's sole purpose is to support the business, although technology can also partially compensate for business weaknesses and therefore be part of a core competency. As a result, business and its IT engage in a supportive feedback loop to build the core competencies in a corporation (Figure 2). New advances in technology have enabled companies to become more agile and more alert to market changes. At the same time, these advances also cause markets and traditional business models to change. It should not be a surprise to

Figure 2. Core competencies of companies

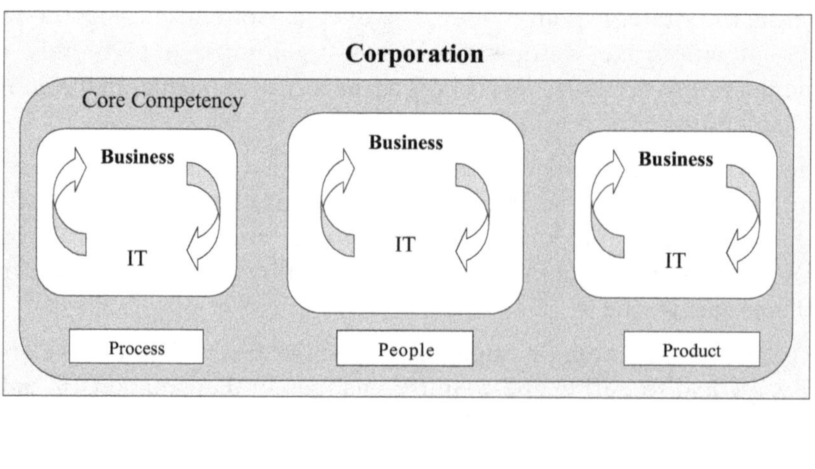

discover that one of the top five concerns CEOs have voiced in surveys since 1997 is how to align business with IT, as Pantaleo disclosed in the preface to this book. These managers have realized that their companies cannot reach their full potential without the support of information technology.

People

People and their intellectual contributions are the premiere asset of a company. The combination of different individuals with their ideas, visions, skills, and capabilities determines a company's personality. People are the ones who have to change, hence they dictate the pace and timing of change. People also enable the IT function of the company and are the ones that have to make the IT work for the business. Technical people with business skills will use technology to create IT solutions that support business. requirements

Processes

Processes ensure smooth operations by providing a framework for the organization. They define the organizational structure and boundaries by controlling the integration of suppliers and customers. Processes evolve over time and are constantly impacted by technological changes. Without processes, companies would have no consistency in their daily operations. This is also the place where continuous change needs to be embedded and the first place to create efficiency. Continuous change and efficiency are both supported by IT, which allows the facilitation of information exchange, and the implementation of best practices through system analysis and design.

Product

Knowledge about the product is a core competency for a company. This knowledge includes information about current and future market developments. Knowledge about the product also serves as an indicator of the effectiveness of the most current best practices. IT is able to support this core competency by integrating suppliers and customers into the organizations. Electronic Data Interchange (EDI) is a technology that allows companies to easily communicate through a set of common standards. New technologies such as XML are built upon this successful concept and will soon replace EDI. Even though, technology by itself cannot be a core competency, it allows organization to strengthen their core competencies and even compensate for weaknesses to some extent.

Figure 3. Production value chain

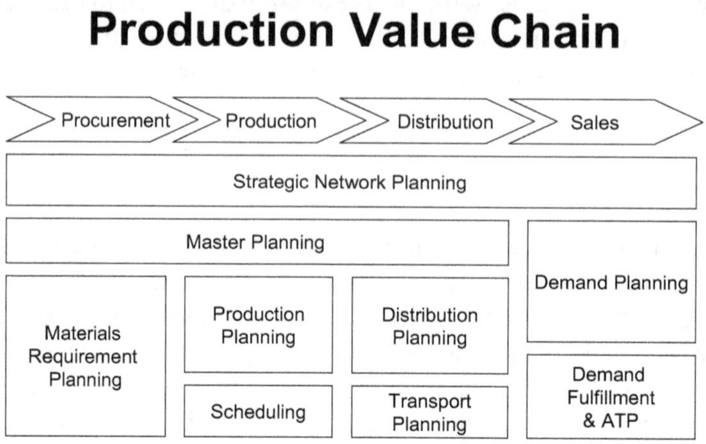

Effective integration among people, processes, and products is essential to ensure that the value chain in the company is integrated in order to provide high value to the customer (Figure 3). The components of the value chain that must be integrated are shown in this figure and are explained next.

Customer Integration: The need to integrate a company's product line with the needs of its global customers has become increasingly important. The functions that should be integrated are: sales support, forecasting demand, generation of orders, order entry, quoting and promising deliveries, demand management, logistics, distribution, and field service (Langenwalter, 2000).

Engineering Integration: About 90% of a product's cost is determined during its engineering design cycle, although its quality characteristics are also determined during product and process design. The functions that need to be integrated here include: the design process, product phaseout, product data management, integration with suppliers, project management, and company-wide integration.

Manufacturing Integration: This includes planning and execution of the materials and manufacturing functions, in addition to the supply network that feeds these activities. The functions to be integrated consist of: materials and capacity planning, enterprise production systems, Just-in-Time management, advanced planning and scheduling, supplier integration, quality management systems, and maintenance operations.

Support Service Integration: Support services are those functions that do not directly touch manufactured items or contact with customers, but are required for a company to operate. These include: accounting, costing, human resources, and environmental management.

Companies are being forced to integrate the above functions seamlessly so as to deliver high-quality products to customers anywhere in the world at any time of the day. The use of on-line ordering and tracking systems heightens customer expectations as the functions of a company become transparent. These changes, in turn, increase the expectations from the IT systems.

Role of IT

An Enterprise Resource Planning (ERP) system is designed to integrate multiple functions within the entire company. It predicts and balances demand and supply. It is an enterprise-wide set of forecasting, planning, and scheduling tools which links customers and suppliers into a complete supply chain, employs proven processes for decision-making, and coordinates sales, marketing, operations, logistics, purchasing, finance, product development, and human resources (Wallace & Kremzar, 2001). Its goals include high levels of customer service, productivity, cost reduction, and inventory turnover, and it provides the foundation for effective supply chain management and e-commerce. It does this by developing plans and schedules so that the right resources — people, materials, machinery, and money — are available in the right amount when needed. When a company decides to implement an ERP system, every part of the company has to adjust in order to integrate the new business tool into their daily operations. As in a smoothly running chronograph, which requires each individual gear ring to be readjusted when adding a new functionality such as moon phases to its system, the company will have to readjust. It will require careful preparation to successfully implement change. During the first step, the current system has to be halted and taken apart, then changes are implemented and, finally, the new structure has to be reinforced and supported to stay in place. This concept is commonly known as the unfreeze-change-refreeze model.

During an ERP implementation, every employee will be affected by the changes in daily operations. Managers often have to completely reorganize the company's policies, such as the compensation scheme or corporate culture. To better understand the difficulties involved in implementing an ERP system, we will discuss organizational change concepts next.

Change Management

In spite of the significant benefits they offer, many ERP systems fail (Stratman & Roth, 1999). Many ERP systems face implementation difficulties due to the resistance of workers. Al-Mashari and Zairi (2000) assert that effective implementation of ERP systems requires five core competencies to be established, among which is the use of change management strategies to promote the infusion of ERP in the workplace (Aladwani, 2001). The following is an excerpt from an October 2003 cover story in *Network Magazine India* (Pereira, 2003):

- Most companies considering ERP today are looking at migrating processes and data, from legacy systems to modern computer systems running an ERP application. Fresh ERP implementations, or 'greenfield' implementations, are rare. And migration brings along a whole set of problems, the primary one being people.
- An ERP implementation impacts people, systems, and the organization as a whole. And barriers are expected from these areas since work processes are expected to be altered when ERP is introduced. Business processes, roles, and responsibilities would undergo change. And when this happens, the organization will encounter a sticky issue called 'resistance to change'.
- Since ERP introduces transparency in operations and brings about more discipline, people who are empowered are bound to oppose it. People who are very used to a particular system or a certain set of processes, won't accept an alternative like ERP so easily and won't adapt to the new system.
- Besides employees, people from the extended enterprise will also protest. Vendors for instance may not take to the idea of updating data in the ERP system through a Web interface. Change is welcome if it reduces workloads or gets the job done faster. But ERP does not offer exactly that.
- ERP consultants say the resistance to change can be minimized through initiatives like change management, periodic training, 'hand-holding,' organization preparedness, and of course, management support.

Change Management Life Cycle

In order to understand the changes involved in implementing an ERP system, it is worth taking some time to learn about the phases involved in change management life cycle (Figure 4). Understanding these phases also serves as a guideline for you to successfully implement change. The change process can be disrupted during any phase of the life cycle, resulting in termination or a need to go back and repeat parts of the process. In this section, we will discuss each stage of the management life cycle.

Figure 4. Change management life cycle

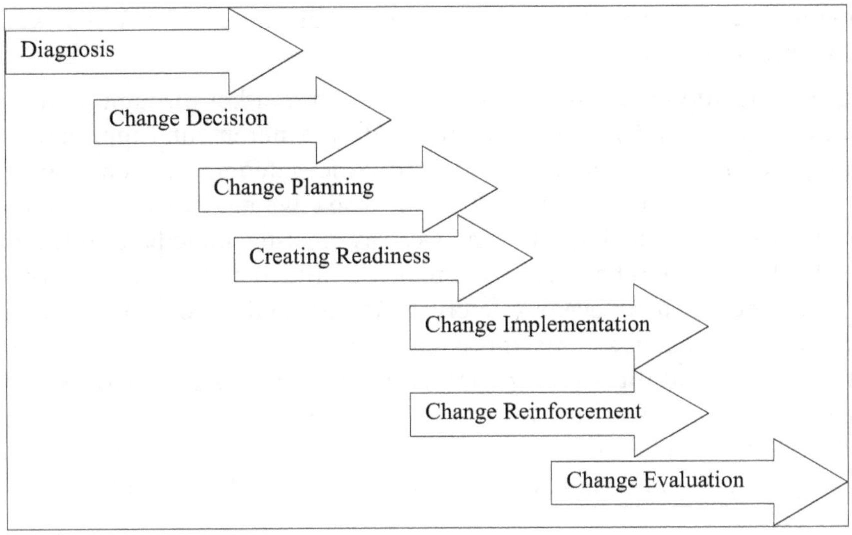

Phase 1: Diagnosis

The initiative for change must be supported by the company's leaders. Hence, the diagnosis of the current system can occur formally, as part of a management plan, or informally, by discovering the need for change during daily operations. In order to assess the need for change, a team has to look at the current position of the company and where the company wants to be in the future. If the gap between present and future is large and important enough to the company, the need for change will be obvious.

Phase 2: Change Decision

There are six major change components within a company, namely people, technology, strategy, culture, structure, and tasks. It is important to understand that these components are interrelated and that a change to any one of them will impact all the other components. For example, people will always be affected by changes made to any of the five components.

The gap discovered during the diagnosis phase can be a result of either reactive or proactive change needs. For example, the Sarbanes-Oxley Act of 2002, which introduced tougher corporate financial disclosure requirements, is an environmental change that forces public companies to be reactive. In contrast, implementing an ERP system might be a proactive change, since the company is anticipating that such a system will be essential for future business and wants to be fully prepared for the new system.

Change can be implemented in two different ways, either through incremental changes or by changing business fundamentals. An incremental approach cycles through several smaller changes, evaluating the outcome of each change and making adjustments to the overall situation. The advantage of such an approach is that it is more visible to the people involved by creating immediate and recurring feedback, allowing the change process to be flexible. In the fundamental change approach, things are handled completely differently and basic assumptions about the company are changed. This approach allows for a more thorough planning process and has the ability to completely transform a business. The disadvantage lies in any changes that occur during the planning phase, which lead to either the need for additional planning or in the worst case rendering the whole process obsolete. This phenomenon is often referred to as "analysis paralysis".

Phase 3: Change Planning

An integrated plan for change must include a comprehensive examination and alignment of all existing policies, procedures, systems, programs and practices, since they directly influence the six major change components.

Phase 4: Create Readiness

The change leaders are the ones that everyone focuses on. Hence, it is important that these individuals "walk the talk" when they lead the change. These individuals become the role model for change and need to align their behavior, attitude, and appearance with the goals of the change process.

Management will be responsible for providing an environment that is conducive to change. This will include restructuring the reward and benefit structure and the selection and promotion process. They will also need to separate themselves from any employees who try to sabotage the change effort.

The group leading the change has to ensure that the corporate environment also falls in line with the new goals. This includes changing rites, symbols, signs and stories, and encouraging new ceremonies that celebrate the change.

Phase 5: Resistance to Change

The problem with most change initiatives is that they are forced upon many of the company's employees, who do not perceive the need for change and will therefore try to resist it. Some individual sources of resistance arise due to a lack of information about the change initiative, fear of the unknown, risk aversion, mistrust, threats to personal well-being, habit, or even just differences in opinion. In addition to the individual sources of resistance, many companies fail to overcome the organizational barriers to change. An organization needs to ensure that its structure, the culture, the values, the resources provided, and the contracts provide a general environment that is conducive to change.

Once the existing situation has been unfrozen and the anticipated change has been implemented, it is important to refreeze the new corporate equilibrium. Refreezing can be achieved by strictly following the new guidelines and reinforcing desired behavior.

Phase 6: Change Evaluation

The most overlooked stage of a change initiative is the evaluation of the change process and its outcome. Reflecting on and evaluating the change allows the company to extend its corporate knowledge base and hence its competitive advantage.

We all know that change is a phenomenon that constantly surrounds us and that we are unable to avoid it. Consequently, resistance to change is often portrayed as a negative trait or circumstance during this introduction to change management.

However, the positive consequences of change resistance include the reduced bureaucracy and improved communication among team members. Resistance also reduces the likelihood that companies and managers will implement unnecessary changes by forcing change agents to answer the following challenging questions:

- What changes are needed? Are the ones proposed appropriate?
- Can I or we successfully make the change?
- Is the organization serious about this change?
- What is my role in the change process?

Change Management Issues in Implementing ERP Systems

Based on the phases of the change management life cycle discussed above, lessons can be learned about how to implement and when to choose not to implement ERP systems. The importance of undergoing a thorough process is highlighted by comparing the theories discussed above with the top three mistakes in implementing ERP systems identified by CIOs (Weightman, 2003):

1. Embarking on the journey without a solid, approved business case, including mechanisms to update the business case continuously and to ensure the savings are baked into operational budgets.
2. Treating this as a technical project vs. a change that balances people, process, and technology; not using the power of the new, integrated information.
3. Trying to create a solution incompatible with the company's culture.

This list of mistakes demonstrates the importance of understanding change management principles and applying them to the successful implementation of an ERP system. Without an exposure to change management principles, managers may not be aware that they will face major difficulties outside of the technical aspects of such a system implementation. Based on practices reported in the literature, we discuss below major change management issues that need to be considered when an ERP system is being implemented in a company (Wallace & Kremzar, 2001).

Phases 1 & 2: Diagnosis and Change Decision – Approval of ERP Projects

In order to get any ERP system project started, it usually has to go through a lengthy approval process. This implies that people with a non-technical background will make the decision about the project's value to the company. Most manager's and decision makers' views of IT are that it serves only one purpose: supporting the business processes to become a more efficient corporation. Hence, most ERP system projects have to prove that their implementation will have a direct effect on the company's profits or, in business terms, show a positive ROI. As we have already discussed, it is very difficult to put an exact dollar amount on the possible benefits of an ERP system, which must be estimated in order to calculate ROI. The principle challenge in evaluating the future cash flows of an ERP system is to directly link cost savings to the new IT system.

Let's assume that a sales manager decides to target a new customer group. As a consequence, sales immediately increase which, in turn, increases the company's net profit. Now the challenge becomes to figure out if the decision was made as a result of implementing an ERP system or whether it was just a brilliant idea of the sales manager that would have happened anyway. The profitable decision might not be a direct result of the ERP system, but might have been a result of the way an ERP system allows for data to be presented. The sales manager's bright idea may have been sparked during the reengineering attempts necessary to efficiently implement an ERP system.

In a different scenario a manager makes the strategic decision to change the company's sales channels. Is it really possible to say that this was due to superior information provided by an ERP system? The implementation of an ERP system will influence every facet of the corporation, making it impossible to precisely attribute improvements to technology or humans. A fully implemented ERP system provides a comprehensive view of the business and its activities, allowing constituents of the system to make more informed decisions.

The problem faced by management is that an ERP system goes beyond merely supporting the business processes of a company. Efficiency by itself can be measured by comparing inputs and outputs of a process before and after the change. However, an ERP system also changes the way business operates, not simply the processes within the business, so you would have to measure the benefits to the whole company both before and after ERP implementation. As you can imagine this is impossible, since there are too many variables outside of the ERP implementation that have an impact on the business' bottom line.

Ultimately, the decision to implement an ERP system comes down to either being necessary to compete within a certain industry, or whether it contributes to building a competitive advantage. In the case of a competitive advantage, it will be very difficult to measure the ROI of the project and the decision will have to be based on the experience and imagination of management.

Phase 3: Change Planning – Initial Education and Training

To successfully implement the ERP system, the company's employees all need to become involved in the project. During this stage, change management efforts play an important role and need to be carefully planned, as well as implemented. Wallace and Kremzar (2001) recommend that at least 80% of the employees should receive education on ERP systems, as they have to understand the technology and usage behind the ERP system. It is important to include everyone and ensure that they understand why, what, and how their job and the organization are going to change. Also, most people have to see some benefit to themselves in order to support change.

This stage will evolve into *ongoing education* and as a whole will last for the duration of the project. During the ongoing education phase, new employees will receive their initial training and employees that have already been trained will receive a refresher course. This is critical for the smooth operation of the ERP system that should already partially be in place. Employees will have the chance to voice their opinions and improve their skills. It is not uncommon that these educational efforts will cause some friction between the project managers and the employees, although listening to the opinions of employees will enable the project managers to improve the ERP system. Nevertheless, every project will encounter people who are not willing to change, and these people can be very dangerous to the success of the overall project.

Phase 4: Creating Readiness – Fear of Technology

As with most IT projects, employees will have to overcome their initial fear of new technology in order to successfully implement the system. Frequently, employees are very reluctant to learn a new technology. It will require intensive training and new incentive programs to motivate employees to adopt a new technology such as an ERP system. This is especially true if they have not had any previous exposure to database systems.

Most people are reluctant to change their habits and venture outside their comfort zone. It is crucial that employees become involved in the ERP implementation and thus feel that they have a stake in its success. Employees should be able to provide input into the analysis and design of the ERP system to give them a feeling of ownership. Gradually, everyone impacted by an ERP system needs to be trained to help them feel comfortable with the new system. In addition to training employees, the company needs to ensure that there is sufficient technical support available for the problems that will inevitably arise. This requires a well thought out plan of communication and support infrastructure.

Phase 5: Change Implementation – Lack of a Champion

Once a project is approved, it becomes crucial for it to have the full support of the company's top management. In many cases, managers will not comprehend the complexity of implementing an ERP system. The view of most non-technical mangers implementing an ERP system is that it is "just another IT project". However, as you will see in the Robert Bosch case study, nothing could be further from the truth. This vague understanding of what it takes to implement an ERP system often makes it difficult to recruit a project champion. Most managers are not eager to back a risky project that they are not 100% confident will be successful. Even if they support the project, individuals will often be reluctant to put their name on the line for a project that has a high potential for failure. Consequently, it is essential that an ERP implementation is backed by the company's leaders. The key to a successful implementation of such a difficult task is the support and teamwork of everyone involved. Unfortunately, teamwork is not a skill many IT professionals view as important and this makes it even more difficult to put together an efficiently communicating team.

Phase 6: Change Reinforcement – Ongoing Process

Any ERP implementation also faces the common misconception that once the technology is in place the project is completed. In fact, the ERP system must continue to change and evolve with the company as a whole. Every time the organization expands through mergers or acquisitions, these new pieces have to be integrated into the company's ERP system. New technology such as data warehousing projects will also have to be integrated into the ERP system. The complexity of an ERP system and the difficulty of working with them stems from the way the system becomes an integral part of the entire organization. Consequently, every change the organization experiences will be reflected in its ERP system, causing the company to repeatedly modify and improve it.

Phase 7: Change Evaluation

Once an ERP system has been successfully implemented in a company, it is common for the company to disband the system team and assume that it will remain successful without the need for further effort (Wallace & Kremzar, 2001). However, in reality companies need to maintain the system and fine-tune it constantly. Measures have to be devised to see how well the system is performing and the results from the measures have to be used to improve the process, product, and people. In addition, new software is constantly being released and these require the expertise of the implementation team to integrate new pieces of software into the company's older systems.

Challenges in Implementing ERP Systems

One of the key success factors for an ERP system in a company is how it is implemented. This issue must always be taken into consideration because besides the technical issues and features of ERP systems, effective implementation requires a great deal of work and coordination. The implementation of the ERP system has to be investigated and the pros and cons carefully weighed in order to get the most out of it. There are several issues that must be taken into consideration before deciding on an ERP system.

The deciding factors can be summarized under three main groups, comprised of structural issues, technical issues and managerial issues. Note that there is no clear line that separates these issues, and they might be closely interconnected.

It is very important that companies do not rush into a decision when opting to implement an ERP system. The issues should be carefully examined and the decision makers must be aware not only of the benefits they may gain, but also of the risks. Obstacles may arise due to structural problems such as reengineering problems during the implementation, as well as the technical problems involved in replacing existing IT architectures with new systems. One thing is very clear: ERP implementation is not simply a technological issue. A good balance between technology and business issues must be achieved in order to reap the benefits of a successful implementation. Next, we will discuss the three main categories of issues that arise and explain them in detail.

Structural Issues

We mentioned earlier that ERP systems help to integrate business processes, thereby supporting the formation of a more disciplined and uniform organization. This is due to the fact that ERP systems help to standardize the processes within the company and thus support uniformity. However, this requires companies to define their business processes clearly and have a broad understanding of business implications. Not every company gains an advantage by standardizing their business processes; some companies' strengths may stem from their successful practice of customized processes. If a company's competitive advantage results from this customization, implementing an ERP system and thereby standardizing their processes could not only lead to a failure in implementing the ERP systems, but could also result in a serious business failure.

Before the ERP era, companies defined their requirements and their business processes and then wrote their own software to support these processes. Since

ERP systems became available, whole procedures have changed. Now, the companies have to redefine their business processes in order to be able to use ERP systems. This means that ERP vendors are defining and writing the business processes for the company based on best practices. But the question that arises is: do a company's processes fit into the predefined, standardized processes of the vendor's software?

This problem might be clearer with an example from the computer industry. Apple, with its distinct product portfolio does not have to worry about the standardization of business processes. Their competitive advantage occurs due to the uniqueness of their products. After implementing ERP, standardization might give them some cost advantages and will not have a negative effect on their competitiveness. On the other hand, for the PC manufacturing companies, standardization could be a real danger because their products are more or less based on the same technology. What differentiates one company from the other is its service and price. Thus, standardizing the business processes by implementing an ERP system might be fatal for a PC manufacturing company. Through standardization, the risk of losing competitive advantage threatens the company (Davenport, 1998, p. 6).

Another structural issue concerns global implementation versus local implementation. This issue is also related to the standardization issue. A company might, for example, be a global company trying to implement an ERP system. By doing so, it may lose its ability to adapt to local conditions, thereby endangering its business. By standardizing its processes, a company can lose its competitive advantage to local, more flexible companies. Thus, before implementing an ERP system, a company should decide whether to implement it globally, or implement some modules globally while adjusting other modules to fit local requirements. Thus, integration may occur without the loss of competitive advantage. However, this kind of implementation is more expensive and takes longer to complete. Everything depends on the business strategy. A company may have global business processes in the logistics area if it does not matter where in the world a specific part is produced. Here, it makes sense to have globally standardized processes. However, typically the payroll system must be localized due to each country's specific taxes and legal requirements.

One possibility is to implement only certain modules, such as finance and logistics, globally. Because ERP systems are based on modules, companies can choose among suitable modules for their company. For instance, a service company might not need PLM (Product Life Cycle Management) systems, or a company with few employees might find it unnecessary to implement an HR (Human Resources) module. This gives companies more flexibility. One issue related to this is customization. Companies can, up to a certain point, customize the system. For instance, it is possible to choose whether revenues are shown regionally or geographically. However, customization does not offer the option of fitting the system to the business. In this case, companies may change or enhance the system to adapt it for their processes, but this is likely to cost more and may take much longer to implement.

An important issue is often the choice of whether to use a single vendor or choose the most appropriate modules from different ERP vendors to support a company's business processes. The single vendor approach enables the company to ensure compatibility and the only partner it has to deal with is the single vendor. In the fragmented approach, also referred to as the best practice approach, one module may be chosen from one ERP vendor, e.g. logistics, and another module from another ERP vendor, e.g. Human Resources. An example from the real world, in this case from Dell, illustrates the differences between these two approaches:

Dell Computers started with the implementation of an ERP system supplied by the largest ERP vendor, SAP, in 1994. Two years into the implementation period, they abandoned the SAP software R/3. Dell's CIO, Terry Kelley, explained their reason for leaving the SAP R/3: "Over the two years we were working with SAP, our business model changed from a worldwide focus to a segmented regional focus... [SAP] was too monolithic to be altered for changing business needs." After this, Dell decided to go with a fragmented approach and thus started to work with i2 Technologies (an SCM provider) to manage raw materials. The order management problem was solved by choosing software from Oracle. One year later, Dell decided to work with Glovia for manufacturing. Kelley said: "We keep carving out pieces of the puzzle and delivering quicker value than if we were putting in one huge ERP system." (Austin, Cotteleer, & Escalle, 2003, p. 4)

Technical Issues

Implementing an ERP system is not simply a technical problem; the software itself costs only 15% of the total cost of the ERP system. This clearly shows that structural and managerial issues should not be underestimated. Table 1 shows the proportion of implementation costs in different areas.

The technical issues are mostly related to structural issues such as customization or writing company-specific software. One difficulty concerning technical issues is the integration problems of different systems. Without ERP, companies use separate systems for each area, one for finance, one for logistics, one for HR and so on. The difficulty is to combine all these different systems to one integrated system. Old, so called legacy systems, must be revised and the data

Table 1. Implementation cost for each area and their range (Mabert, Soni, & Venkatarmanan, 2001, p. 75)

Area	Average Cost	Range
Consulting	30%	20-60%
Hardware/Infrastructure	25%	0-50%
Implementation Team	15%	5-20%
Training	15%	10-20%
Software	15%	10-20%

has to be transferred to the new system. This is likely to take a long time and cost a great deal of money.

Another problem occurs when companies go on using their old legacy systems for some modules and implement the new modules as add-ons. In this case, there must be interfaces so that the systems can talk to each other. Creating these interfaces is a very difficult and time consuming problem, which only becomes greater over time due to the need to maintain all the interfaces as the different systems change.

The degree of these technical challenges depends on many different product characteristics. One is the complexity of the ERP system. Typically, these systems cover very different kinds of businesses so the number of parameters and their interdependence could be extremely high. Combined with this there is a huge amount of knowledge required and most companies are not able to build up this knowledge in-house so they must rely on the support of outside consultants. However, these consultants will not be familiar with the company communication issues that arise. The second is the question of the technical standards used by the vendor of the ERP systems. If a software system is based on proprietary technologies and interfaces it is much more difficult to integrate the ERP software into an existing application portfolio. It could thus be beneficial if the software is based on industry standards such as J2EE, .NET, etc. These will be discussed in the next chapter.

Managerial Issues

Managerial support is very important for the implementation phase of an ERP system. As mentioned earlier, effective implementation is not only a technological but also a managerial issue. The change to ERP must be strongly supported at the managerial level and must be controlled at all levels. As it can be seen from

Table 1, the consulting, training and implementation team make up more than half of the total cost, and these are all managerial issues.

In large companies, the implementation period is lengthy, possibly as long as several years. Thus, it is important to work with the appropriate implementation teams. Together with consultants from the ERP vendors, the implementation team plays a major role in the implementation. It is important to choose the right staff for the implementation. For instance, Cisco chose their most skilled employees from several departments for the implementation team and they worked together closely to ensure a successful implementation. Thanks to this successful orientation and combined with other success factors, Cisco was able to implement an ERP system in only 9 months, an impressive feat for such a large company (Cotteleer, Nolan, & Austin, 1999, p. 11).

Another problem that must be addressed is the reaction of users to the new system. They might react adversely to the system, hindering its success by resisting using it efficiently. Thus, a successful training program accompanied by an incentive systems to encourage them to use the new system is very important. One of the reasons for the failure of many ERP systems is the resistance of potential users. This problem should not be underestimated by managers and appropriate steps should be taken. One of the key success factors is the support for the project provided by the company's top management itself. If the top management strongly backs the project team, they will find it easier to convince doubters.

The main difficulty from the managerial perspective is the difficulty of estimating the budget and the schedule. It is almost always underestimated, and budget and schedule overruns are one of the main reasons why ERP implementations fail. Thus, a realistic estimation of both the budget and schedule reduces unexpected overruns at the end and consequently also reduces the frustration.

Implementation Strategies for ERP Systems

Phases

The implementation of ERP systems is normally implemented in a phased manner. The implementation approach typically entails four phases, although this may vary depending on the company's preferences.

The first step is the planning phase. The scope of the project is defined in this phase. Detailed activities are planned, the concepts are established and the technical requirements are set up in this phase. The planning phase is important and should be finished completely before moving on to the next phases, because failures that occur once the project implementation is underway are far more difficult to correct.

The second step is the realization phase. The purpose of this phase is the customization and configuration of the system. The activities in this phase include creating interfaces and conversion programs, creating the documentation required by the company, and setting up authorizations.

The third step is the preparation phase. This is the phase where the system is readied by the project team for its future users. The steps in this phase typically include the customisation of the system, integration and quality assessments, and conversion of data from the old system to the new system. The users should be trained to a high standard and supported with adequate documentation, so that after the handover the system can be used smoothly.

The last step is the production phase. The system goes live in this phase and is handed off to the prospective users, who assume full responsibility for its operation. The project team's role now becomes to support the system while at the same time fine tuning it (Welti, 1999, pp. 5-7).

If properly implemented, these phases help to achieve a successful implementation and a smooth transition from the old system to the new one.

Strategies

The implementation strategy defines how the implementation should be realized. There are three possible implementation strategies: step-by-step, big bang, and roll out.

Step-by-step implementation is a strategy of implementing the system in small steps. Not every module is implemented at once, and usually a single module is implemented at a time. The complexity of implementation is reduced, since the effort required for the coordination, control and organization of the project are reduced. The costs are spread over time, making it affordable for many companies. Also, with every step the project team gains knowledge which they can use in the subsequent implementations. In this strategy, however, many interfaces have to be built so that the new modules and existing modules can communicate with each other. This implementation strategy stretches over a long period. The creation of interfaces is not always easy and can be a cumbersome job. Also, the motivation of the project team decreases towards the end of implementation and the knowledge gained with every step may be used to go back and redesign previous steps, which prolongs the implementation further and makes it more difficult.

The big bang implementation strategy replaces the existing system in a single operation. This strategy entails an intensive testing phase to check business processes in detail. There is no necessity for creating interfaces in this strategy. The motivation of the project team is generally high and the time necessary for a big bang implementation is relatively short. On the other hand, the implementation is complex and the employees are under considerable stress during the implementation. The workload involved in training all the users at once before the system will be used is huge. A high degree of consultant support is necessary in this strategy.

The roll-out implementation strategy creates a model implementation for one area of the company and then rolls it out to other areas. There is less risk here because most of the problems are already solved in the model implementation. This strategy ensures that in all the divisions of the company, uniform systems are implemented. The experience gained in the model implementation and the accompanying expertise makes a fast implementation possible. The problem here may be that in some divisions specific processes can be overlooked. This roll-out strategy is typically used by huge multinational companies. Often, a model implementation starts in the company's home country. The roll-out process must be adjusted to take into account the specific requirements of the different environments. In particular, more localization might be needed in some countries due to their different business practices and legal requirements (Welti, 1999, pp.7-9).

Chapter Summary

Conducting a change is more complex and difficult than simply imposing desired behavior on individuals. It is necessary to carefully plan how to unfreeze current behavior, change, and refreeze the desired changes. A change agent needs to convincingly present the change "message", create readiness for change, implement the change, reinforce the change and evaluate the change. Every company will experience different problems and difficulties trying to implement change and may be forced to terminate their change initiative during any phase of the change life cycle. Due to the scope of a comprehensive ERP system, the entire company needs to be involved in the process. Hence, the aspect of changing individual behavior and creating change readiness occurs in addition to the technical difficulties in implementing an ERP system. Such a major change effort requires that the company's management understands and fully supports the project. Once the rational for supporting an ERP implementation is understood and accepted, approximately 80% of the money and effort required will be devoted to creating an environment that is receptive to change. Some of the additional problems that are involved in creating an environment that is receptive to change for an ERP implementation include the fear of technology and the inherent complexity of the process.

Hurricane warning systems, satellite technologies, electronic Web sites, radio, and TV have all helped many people in the U.S. to escape the wrath of hurricanes, whereas people in developing countries have yet to see the benefits of such systems in their regions. Even with the availability of these systems, lack of effective leadership in implementation can lead to disasters such as that experienced by the people who got impacted by Hurricane Katrina in New Orleans during August 2005. This event showed the administration did not plan effectively for a change that might occur if most of the city got flooded. Even if there was a plan, it was not implemented and coordination at a national level was minimal for the first few days of the disaster. This shows the need for understanding the management issues that are discussed in this chapter.

In time, these technologies will be available to people all over the world, making humanity much safer in the face of natural calamities. Similarly, ERP systems are beginning to play a major role in standardizing and simplifying business processes in large companies and the benefits are beginning to accrue to these companies. Still, the technology is in its early stages, and there are many reported cases of failures in implementation. People in developing countries are not exposed to the utility of such systems and are still used to a large number of people selling and maintaining small retail stores. As the benefit of ERP systems spread further, the economics of scale are likely to spread around the world, allowing people to buy high-quality goods at a reasonable price. Appendix A provides an example of a future ERP system through which excess food produced by American farmers could be used feed hungry people around the world. Similarly ERP systems could be used to make the supply chain much more efficient on a worldwide scale, thereby benefiting people everywhere. As discussed in this chapter, it is critical that people be aware of the need for change and become champions who are ready to lead the change in their companies and countries. Readers such as yourselves can become better informed about availability of ERP system and combine them with good leadership and implementation strategies so that human suffering can be minimized.

We will discuss the technical issues involved in developing and implementing ERP systems in the next chapter.

Team Assignments and Discussion Questions

1. Find an example of a company that has experienced difficulties in implementing an ERP system and relate their problems to the Change Management Framework.
2. Relate some event from your personal experience to the Change Management Framework.
3. Research some additional information about ERP implementations, such as the cost in relation to the size of the organization or the time involved in planning, designing, and implementing the ERP system.
4. Who should be involved in an ERP system implementation? Please identify the importance of these individuals during the different stages.
5. Find examples of other frameworks for change management.
6. Why did it take so long to persuade companies to implement ERP systems?

References

2020software. (n.d.). Top solutions from SAP America, Inc. Retrieved March, 18, 2005, from http://www.2020software.com/software

Aladwani, A. M. (2001). Change management strategies for successful ERP implementation., *Business Process Management Journal, 7*(3), 266-275.

Al-Mashari, M. & Zairi, M. (2000). Information and business process equality: The case of SAP R/3 implementation. *Electronic Journal on Information Systems in Developing Countries, 2.* Retrieved September 7, 2005, from http:/new.ejisdc.org/ojs/

Austin, R.D., Cotteleer, M. J. & Escalle, C. X. (2003). Enterprise resource planning, *Harvard Business School*, 1-8.

Awad, E. & Ghaziri, H. (2004). *Knowledge management.* Upper Saddle River, NJ: Pearson Prentice Hall.

Cotteleer, M. J, Nolan, R. L. & Austin, R.D. (1999). *Cisco Systems, Inc.: Implementing ERP. HBS case study.* Harvard Business School Publishing.

Coy, P & Foust, D. (2005, September 12). Katrina's Wake. *Businessweek 3950.*

Davenport T. H. (1998). Putting the enterprise into the enterprise system. *Harvard Business Review OnPoint Article*, 1-11.

Langenwalter, G. A. (2000). *Enterprise resource planning and beyond: Integrating your entire organization*. New York: St. Lucie Press.

Mabert, V. A., Soni, A. & Venkataramanan, M.A.(2001). Enterprise resource planning: Common myths vs. evolving reality. *Business Horizons*, 71-78.

Made2Manage Systems Inc. (n.d.). Enterprise resource planning overview. Retrieved April 23, 2005, from http://www.made2manage.com/

Major Storms (2005). Retrieved July 11, 2005, from www.factmonster.com/ipka/A0001441.html

Parker, M. M. (1996). *Strategic transformation and information technology*. Upper Saddle River, NJ: Prentice Hall.

Pereira, B (2003). ERP: Striking it right. *Network Magazine India*. Retrieved September 8, 2005, from http://www.networkmagazineindia.com/200310/coverstory01.shtml

Porter, M. E. (1986). *Competition in global industries*. Boston: Harvard Business School Press.

Rau, K. H., Raju, P.K. & Sankar, C. (2005). *Information systems in enterprises: A case study approach* (3rd edition). Anderson, SC: Tavenner Publishing Company

Recklies, D. (2001). Porters 5 forces. Retrieved September 8, 2005, from http://www.themanager.org/Models/p5f.htm

Sarbanes-Oxley Act (2002). U.S. Bill HR3763. Retrieved September 5, 2005, from http://news.findlaw.com/hdocs/docs/gwbush/sarbanesoxley072302.pdf

Smith, A. M. (2002, April). Data warehousing and ERP—A combination of forces. *The Data Administration Newsletter*. Retrieved March, 25, 2005, from http://www.tdan.com

Stratman, J. & Roth, A. (1999, November 20-23). Enterprise resource planning competence: A model, propositions and pre-test, design-stage scale development. *30th DSI Proceedings*, New Orleans, LA (pp. 1199-1201).

Wallace, T. F., & Kremzar, M. H. (2001). *ERP: making it happen*. New York: John Wiley & Sons.

Weightman, C. (2003). The top 10 ERP mistakes (Part 2). Retrieved September 8, 2005, from http://www.sapinfo.net/index.php4?ACTION=noframe&url=http://www.sapinfo.net/public/en/article.php4/Article-279253ea93472d35c8/en

Welti, N. (1999). *Successful SAP R/3 implementation, practical management of ERP projects*. London: Addison-Wesley Pearson.

Appendix A

This appendix was developed by Kelly Hicks, an MBA student at Auburn University, May 2005, and used with her permission. (hickskm@auburn.edu)

World Food System: Proposed Project Description

International aid organizations such as the United Nations World Food Program (UNWFP) and World Vision try daily to address the problem of hunger in the world. We have an excess of food being produced by American farmers, which could be used to feed hungry people elsewhere. However, we have neither the logistics nor the communication systems that would be needed to supply the hungry with our excess food. The government of the U.S., in the form of the USDA, pays farmers to not grow food in their fields to keep the commodity markets from being flooded. This inflates the cost of food and keeps U.S. farmers from going out of business. However, if there was a way to use information technology to increase communication and information in agriculture, we could use the U.S. farmers' ability to produce food to feed people internationally. This worldwide issue will require expert design in logistics architecture to facilitate the distribution of food from the US to the international aid organizations.

At this point in time, the UNWFP and World Vision partner to purchase food, which is then shipped to needy areas. The UNWFP and World Vision already have logistics and technology in place to support supplying remote areas with food. Their technological infrastructure is supported globally. Currently, they purchase food from many different vendors internationally to supply needy areas. The finance wing of these organizations is always busy with procurement. What they need is a steady, reliable, source of food to supply their organization. The USDA, or some non-profit organization such as a wing of the UNWFP, could facilitate the international exchange of excess food by American farmers.

What this would do for American farmers is give them the opportunity to produce to the best of their ability. The government is subsidizing US farmers today. We would continue to subsidize them, but encourage them to produce as much as possible. The biggest financial contributors to the UNWFP are governments. Instead of giving them money to purchase food, or in addition to it, we can give them food to supply their efforts. This would cut back on the costs for UNWFP associated with paying market prices for food. The total cost of procurement for the agency will be less over time.

After farmers have grown their products, we would then set up communication between CO- OPs, individual farmers, railroads, freighters, air transport, and the international aid organization that we would supply (UNWFP). Once we get our product to the UNWFP, they have the built in logistics to supply the hungry around the world.

This solution not only contributes to the largest humanitarian effort to end hunger, it solves a fundamental problem in American agriculture.

IT Systems and Support

If the system was developed today there are several technologies that I believe would be the basis for starting this project. Future systems could build on these basic ideas.

The IT system needed for such a complicated endeavor will have to be flexible in design, making use of remote technology via satellite, and able to handle international exchange rates, logistics, HR, large volumes of information, and have rapid response time. It must be simple enough for end users to input data, and complex enough to schedule deliveries internationally. Several components will need to be included in this system. The following will all need to be integrated into the overall design of the World Food System.

Enterprise Resource Software must be used to integrate all the demands placed on the system. It provides security, reliability, and scalability, so we could start in one part of the globe and move to others as the successes build. It also allows for data transfer across platforms, which will be integral when dealing with the aid organizations and all the different systems they use.

ERP, or Enterprise Resource Planning, is part of a group of products that aim to reduce supply chain inefficiencies and create opportunities for growth through integrating all functions into one system. ERP is mainly used in an industrial setting. Companies that have successfully implemented an ERP system include Briggs and Stratton and Robert Bosch Corporation (Rau, Raju, & Sankar, 2005). These companies have international components and are very large. They have both implemented SAP, which is an ERP product offered by a specific vendor. There are other ERP systems on the market, such as IBM ERP, MAS 500, and SYS PRO ERP. All of these ERP systems are intended for companies that have annual revenues greater than $50 million (2020software, n.d.). I am assuming that the system needed will be large in order to be applied all over the world.

ERP is designed to vertically integrate all company functions to allow for easy exchange of real time information in the back office regarding the manufacturing process (Made2Manage Systems Inc., n.d.). ERP software usually has a central database as its hub, (Smith, 2002) as do most knowledge management systems. This allows application components to share and reuse data efficiently. The software is divided into functional areas of operation and can be integrated, including interfacing with old legacy systems. Some of the components available are: finance and accounting, material management, customer order management, logistics, and production management. Finance and accounting allows managers to manage cash flows, key accounting tasks, budget forecasting, payroll and HR. Production management allows managers to create job orders, capture costs and monitor production to allow for building products more profitably. Customer order management generates sales orders, allows for custom-built products, and generates quotes. Materials management helps to automate shop floor and inventory data collection. It also allows for tracking materials throughout the production process, which helps manage resources and inventory seamlessly (Made2Manage Systems Inc., n.d.). The logistics component supports complex business processes in areas such as procurement, fulfillment, warehousing. It gives access to real time information and visibility into multiple supply chains. This allows for efficient use of resources.

Figure 5. Schematic of world food system

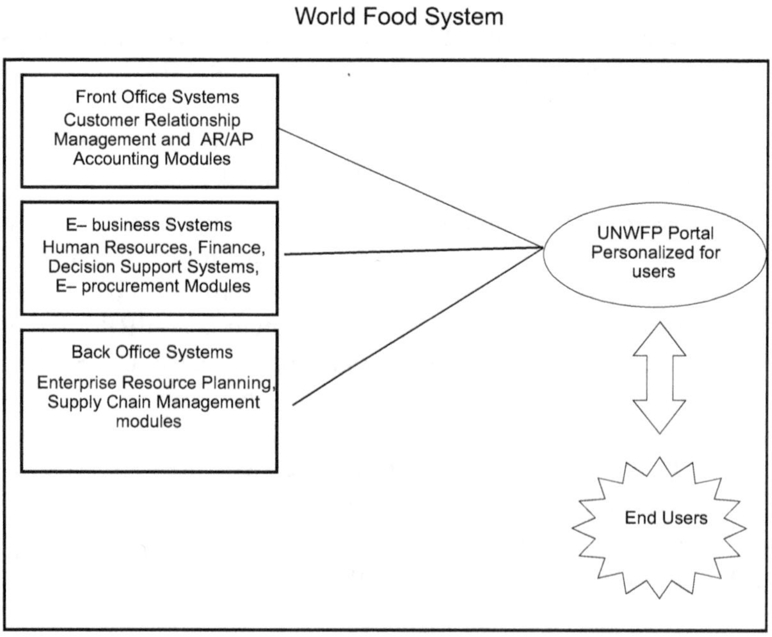

Systems other than ERP will be an integral part of the process, as well as making the system as user friendly as possible. I see the World Food System and its components as being accessible through a personalized portal for every user (Figure 5).

As shown in the graphic, there are many different levels of systems. Each has its own function for an organization. Ours will need to simultaneously operate many systems. This is done by designing systems that work together or by using a specific vendor such as SAP and using all their modules. A module is a specific system such as CRM or a HR component. Which modules are used often depends on the size of the company and the amount that the company wants to spend on the systems. I expect to have all design done in house for the World Food System.

Components for the system include CRM, SCM, and ERP, or a future representation of the systems. These components are explained below.

CRM, or Customer Relationship Management, is a front office system that helps to profile customers for target marketing, make staffing level predictions, make customer profitability, churn, and retention profiles. All of these things help a company to make better marketing decisions at a lower cost (Awad & Ghaziri, 2004). For this system, they can help keep countries and distribution customers separate and honest.

SCM, or Supply Chain Management, is a back office application that allows for interaction between the company, it suppliers, and retailers. It manages JIT inventory and keeps all companies operating as efficiently as possible to control warehousing costs. Companies such as Wal-Mart and Dell are pioneers in this field in the U.S. All parties are usually linked though an extranet via the Internet. Security is a primary concern and firewalls will be in place to make sure that suppliers only see the information they need. Due to the short life-span of many of the products, we will need the system to be designed correctly to keep the food from going bad.

These are just a few of the applications available that we can build our system on. The future may hold changes and a variety of new ways to make this process more efficient. Staying on top of future technologies and systems is important to making this process as easy as possible. All of these applications come with a cost and we may have to decide which one can give us the most bang for our buck because our budget will be limited.

The size of this project would require in depth analysis of all logistics software and the IT backbone that will support it. The idea behind the project is to use IT and IT architecture to facilitate timely deliveries to areas around the world.

We can choose what kind of legacy we would like to leave behind when we pass on. We can either work to make the world a better place and put more into the future of everyone or we can take from the world and let people suffer. I choose to be a part of the solution. Our world is rapidly getting smaller and if we don't make the effort to treat everyone with the respect and dignity they deserve we are dooming the future. This includes all people, plants and animals on earth. We can restore ecological balance and integrity to those who have suffered. This is not an easy undertaking and requires experts at every level fighting for those who are less fortunate.

My desire to develop this system to help hungry people and help save the American farm comes from my desire to leave the world as a better place than I have found it. This system is going to be complicated and hard to develop, but since there are others, such as the UN and World Vision, who are fighting this same fight right now, I believe it is possible.

Chapter V

Technical Issues in Implementing ERP Systems

Learning Outcomes

- Understand the fundamentals of relational database management systems
- Be able to define technical terms such as single instance and normalization
- Be aware of how database considerations impact ERP implementations
- Have a general overview of the technical issues in implementing ERP systems
- Become familiar with implementation issues and strategies for ERP systems

Overview

Every ERP system draws upon a vast set of data that describes the entities and relationships in a corporation. The data may reside either in a single database or multiple databases connected to each other using telecommunications networks. The manner in which the entities are structured and how the relationships among them are expressed define the architecture of the database. The architecture, the networks that connect the databases, and the application software that is used to query the database define how quickly and accurately data can be retrieved. This is not very different from organizing a house. In an organized house, it is possible to find items you are looking for very quickly; wasteful additional expenditure is not necessary as there is no need to purchase duplicates to replace things you have lost. However, we all know that it takes good organizational skills to properly store the items in a house. In addition, if members of the household do not obey the rules and store items in the wrong places, then it becomes difficult to retrieve them. An ERP system requires a similar discipline from members of an organization. The data in the organization must be stored efficiently, not duplicated, and should be available for retrieval quickly. In this chapter, we will discuss the technical issues involved in storing the data in an organization and discuss the concepts that form the basis of database technology. In addition, we will discuss the technical issues that often arise when implementing an ERP system.

Database

As we discussed in an earlier chapter, most industries are experiencing a greatly increased pace of change that requires technology, human, and business structures to adapt more rapidly than ever before. Companies need to have the ability to adapt to a changing environment as quickly as possible, which requires that an ERP system be scalable (built in different modules and combined together seamlessly) and flexible (capable of being modified quickly and efficiently). However, it is a major challenge to build ERP systems that can meet a company's business needs and at the same be scalable and flexible. Even though these attributes are highly important in today's business environment, it is difficult to build and maintain systems that can adapt to the changes (French, Bell, & Zawacki, 2000). Therefore, it is important that you learn the basics of a database system so that your expectations for the system can be realistic.

A *database*, by definition, is a collection of information stored in a computer in a systematic way, such that a computer program (e.g., ERP software) can query it to answer questions. The software used to manage and query a database is known as a database management system (DBMS). The purpose of a database is to efficiently store a collection of data.

Figure 1. A typical "file processing system" (APWINC, 2002)

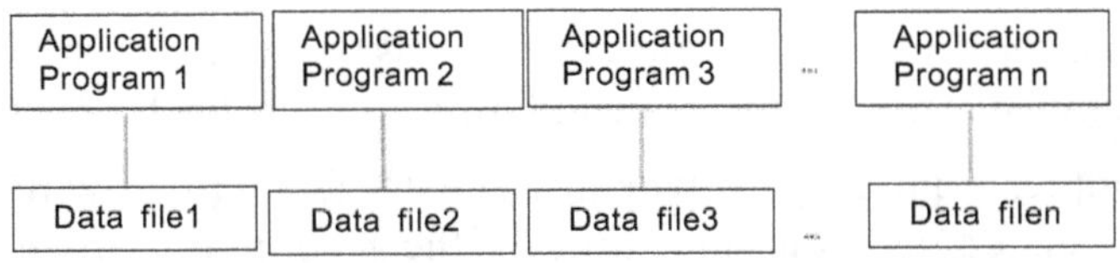

Need for Database Management Systems

The best way to understand the general nature and characteristics of databases is to examine the characteristics of systems that predate the advent of database technology. These early business information systems stored groups of records in separate files and were called *file-processing* systems. For example, a business could have two file processing systems, one for its customers, and one for its customer representatives. Figure 1 depicts the structure of a typical file processing system, where each application uses its own file to store data. Each data file is independent of all the others and is directly linked to the application that interact with it.

Even today, individuals might use such systems to store information about their assets and bank accounts. However, such a file processing system is no longer sufficient for business use, leading to inefficient operations, poor customer relations, and expensive products. Many "mom-and-pop" stores suffer from such difficulties and people must work a lot harder in order to compensate for the inefficiencies of the system.

Over a long period of time, files may be kept in different formats, and application programs may be in different programming languages. Although file-processing systems are a great improvement over manual record-keeping systems, they suffer from major limitations:

- *Data Redundancy and Inconsistency*: The same information may be duplicated in several places, and all the copies may not have been updated properly.
- *Difficulty in Accessing Data*: It may be necessary to write a new application program to satisfy an unusual request.

  For example, to find all the customers with the same postal code, a user might have to access records from multiple files and applications.

- *Data Isolation*: Data is kept in different files and in different formats, so it is difficult to write new application programs.
- *Multiple Users*: When multiple users access the same file at the same time, there might be inconsistencies in updating records, leading to inaccuracies.

  For example, what happens when two customers withdraw funds from the same bank account at the same time? If the account initially has $500 in it, and they withdraw $100 and $50, the resulting balance could be $350, $400 or $450 if no protection is implemented.
- *Security Problems*: Every user of the system should be able to access only the data they are authorized to use.

  For example, payroll can only handle employee records, and have no access to customer accounts, while tellers can only access account data and cannot modify the addresses of the customers. This is hard to enforce using application programs.
- *Integrity Problems*: It is difficult to satisfy constraints.

  For example, no account balance should be allowed to fall below $25.00. If accounts are held in different files, then the application program must be run on all the files.

It is difficult to enforce or to change constraints using a file-processing approach. In order to mitigate these problems, database management systems were developed and have became widely successful since the 1970s (APWINC, 2002). Even now, however, files created using popular spreadsheet programs use a file-processing approach and are subject to the limitations discussed above. In contrast, products such as Microsoft's Access™ and Borland's Paradox™ for Windows are true database management systems.

Components of a Database Management System

Most database systems have a three component structure, as shown in Figure 2. The data itself is stored in a database, and a database management system (DBMS) works with this either directly or indirectly via an application program that allows communication between the user and the database.

A database contains many data elements, such as a customer element, bank element etc. Systems that collect data are called application systems and include point of sales systems (POS), customer relationship management systems (CRM), production systems, accounting systems, order entry systems, and many more. Once the data are collected, they are stored in the database. A database

Figure 2. Database system structure

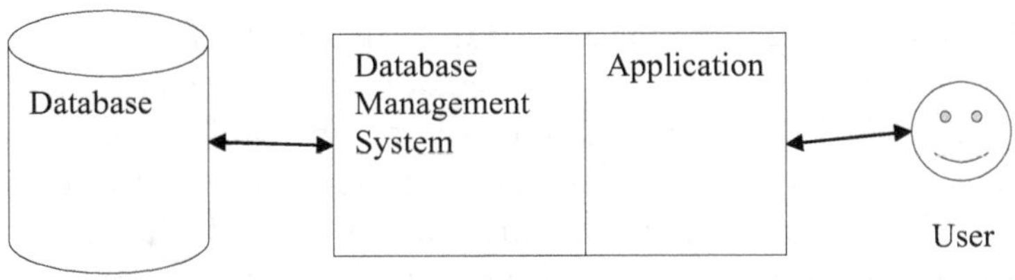

management system is an intermediary that connects the applications and database so that many of the problems that arise in implementing file-processing systems are eliminated and data can be quickly retrieved.

Database Schemas

The data may be structured in a number of ways, known as data schemas. Competing database schemas were proposed during the 1970s, namely the hierarchical model, network model, and relational model. Some systems using the hierarchical and network models are still in use today in data centers with high data volume needs or where existing systems are so complex it would be cost prohibitive to migrate to systems employing the relational model (Wikipedia, 2006). Appendix A at the end of this chapter provides an example of the research conducted by one of the authors during 1970-1980s in order to remove data redundancy from forms. It is surprising that this problem has not been fully solved even now.

Probably the most commonly used data schema is the *Relational Schema.* This is based on mathematical theory and was developed by Dr. E.F. Codd in 1970. Dr. Codd defined only three degrees of normalization in his original work, but today there are several more known normalization forms, although it is rare that any database goes beyond the third normal form. The relational data schema's main purpose is to reduce the quantity of data elements that need to be stored in a database and consequently ensure data integrity and limit redundancy. Defined sets of rules exist for implementing the various degrees of normalization in a relational data schema. This schema incorporates a great number of tables, which are required for the normalization, and is typically used for large data warehouses with significant data requirements (Singh, 1998).

Relational Database

A relational database organizes data into tables that are "related" to one another. For example, one table might contain a customer's information and a second table might contain information about orders. The tables are related to one another because each customer has placed certain orders, and each order is owned by an individual customer. Similarly, an auto parts dealer may store a table of models of cars and a second table of car parts. Each part can be in one or more cars, and each car uses many parts.

Every table in a relational database is organized into rows, where each row represents a single record. The rows are organized into columns. All the rows in a table have the same column structure. For example, the Customer table might have columns for the CustomerID, the ID of the customer representative, the date a transaction took place, the status of the transaction, and other related information.

To understand the issues involved in table design, let us continue with the customer example. We need to know which customer representative is in charge of each individual customer, and as a supervisor it would be very useful to know the customer representative's email address, phone number, and other identifying information as well. A form recording the details about a given customer service might look like the example shown in Table 1.

It is possible to store the personal information of the customer representative with the information about each customer, but that would be very inefficient. If John Doe, a customer representative, has 100 customers, using this structure John Doe's email address and phone number has to be repeated 100 times. Such repetition makes data maintenance difficult. If John Doe changes his email address and phone number, corresponding changes have to be made in 100 different places. Instead, it is possible to replace Table 1 with a unique table called customer representatives, in which each row represents a single person. In the CustomerRepresentative table, there will be a column for the CustRepID. Each representative will have a unique ID, and that field will be marked as the primary key for the CustomerRepresentative table. A primary key is the column or combination of columns that uniquely identifies a record in a given table (Table 2).

Table 1, Example of a customer information table

Customer Rep.	Email	Phone Number	Customer

Table 2. Example of customer representative table

Customer Representative ID	Name	E-mail	Phone Number

Table 3. Example of customer table

Customer ID	Name	E-Mail	Phone Number

Table 4. Interaction table between customer representative and customer

Customer Representative ID	Customer ID	Date and time of interaction	Problem discussed

In order to include all the information shown in Table 1, along with additional useful data, the customer table has be modified to develop two tables, one to store information about a customer and the other to store information such as when the customer and customer representative worked together to discuss a specific problem (Tables 3 and 4).

When answering a query from a customer in a follow-up phone call, the operator can look up the customer information in Table 3 and then the information about the earlier interaction with a different representative from Table 4, and thus be able to provide more effective assistance to resolve the issue. Finally, information on the current phone call will be recorded in Table 4, thereby providing a complete record of all the interactions the customer has had with multiple customer representatives while resolving a specific problem. By "factoring out" the attributes of the representative's address into a CustomerRepresentative table, the redundant information in each customer table is minimized. This process of reducing redundant information from the tables is called normalization.

Normalization

Normalization not only makes the database more efficient, it reduces the likelihood of anomalies. For example, if the customer representative's e-mail address was entered in both the CustomerRepresentative and interaction table (Tables 2 and 4), there is a risk that changing an e-mail address in one table might not be reflected in the other. If the representative's email address was changed in the CustomerRepresentative table, that change might not be reflected in every row in the interaction table. By keeping only the CustRepID in the interaction table, the email address or other personal information in the CustomerRepresentative table can be changed, and the change will automatically be reflected for each customer. In an industry that has high employee turnover, such a normalized data structure ensures data integrity ("Relational Database", 2004, pp. 953-957). The techniques for preventing anomalies are called normal forms and in 1981, R. Fagin defined a new normal form called the Domain/key normal form (DK/NF). He showed that a table in DK/NF form is free of all modification anomalies (Fagin, 1981). If a relationship can be put in a DK/NF, then it will have no anomalies. The trick is knowing how to put relations in DK/NF. As in the example above, it becomes essential to document the data elements and relationships among them. The discipline of data modeling has evolved to address this need.

Data Models

There are two main types of data models, logical and physical. A physical model is used to design the internal schema of the database. It depicts the data tables, the data columns of the tables, and the relationships between the tables. A logical model is a graphical representation of the information requirements of a business area (Carkenord, 2002). A logical data model is independent of the type of physical data storage device used.

One popular logical data model is known as the Entity-Relationship (E-R) model (Chen, 1976). This model provides a picture of the pieces of information that are necessary to run a business. The components of the E-R model are entities, relationships, and attributes. Each entity represents a set of persons, things, or concepts about which the business needs information. Each relationship represents an association between two entities. Each attribute is a characteristic or piece of information that further describes an entity. A name and a textual definition describe each of these components. These names and definitions provide ongoing documentation of the business rules and information requirements of the business area.

Differences between Logical and Physical Models

People often use the words entity and table interchangeably. However, the distinction is significant. An entity represents a set of persons, things, or concepts that are important to the business. A table is a data structure that contains data values and is physically stored on a disk using a specific DBMS. The difference is the link to technology. An entity as such has no link to the technology, while a table is an integral part of the technology.

Why Build a Logical Data Model?

The most important reason for building a logical data model is to document the understanding of users about the business. This ensures that the model meets the business needs. The process of developing a logical model allows the business to drive the system, instead of the system driving the business. By discussing and reviewing the model, extra requirements and problem areas are revealed and can be dealt with early in the development process.

A logical data model is the foundation for designing a database that will support the business requirements. It allows designers to start with a complete picture of the business requirements and then determine the best approach for implementation. It also promotes data re-use and sharing, since the data stored in databases is stable over time. Building a logical model also decreases system development, maintenance time, and costs. Design, coding, testing, and implementation proceed far more smoothly once the business requirements have been clearly identified.

Example of an E-R Model

An example of an E-R model that is being used with a Knowledge Sharing System currently being developed at Auburn University (www.eng.auburn.edu/aetap) is shown in Figure 3. This model documents the structure of a community of practice that is being developed in Alabama to disseminate and transfer engineering technology from academic and research institutions to the industrial community. The community will link university students and faculty with state agent liaisons and contacts within industry in both a personal and virtual manner. The virtual component of the community will be centered on a Knowledge Sharing System (KSS) that will serve as a communication, data collection, and knowledge management (KM) forum. The goal of the community and the KSS is to bring together partners so that they can learn from each others' experiences and avoid having to "reinvent the wheel."

Figure 3. Example of an E-R diagram

Each rectangle in Figure 3 is an entity and the arrows among the entities show the relationship among them. For example, the relationship between a client and a project is shown by the intersection entity, "serviced_client." It took more than two years before the system development team was able to agree that this E-R diagram adequately depicts the information flow and interaction among the partners in this project. The project team developed and discarded more than five versions of the E-R diagrams before this design was selected for the database system. The project team developed a detailed description of each entity shown in this figure. For example, Table 5 shows the details of the attributes of the entity "client." The documentation for this project includes definitions of all the entities involved, their relationships, and descriptions of the attributes for each table.

Relational Operations

In order to produce reports and use a database effectively, we can request data from a relational database by sending it a query that is written in a special language, usually a dialect of SQL (Structured Query Language). Although SQL was originally intended for end-users, it is much more common for SQL queries to be embedded into software that provides an easier user interface. Many Web sites perform SQL queries when generating reports.

Table 5. Attributes of the entity "client"

COLUMN NAME	DATA TYPE	DATA LENGTH	NULLABLE	DATA DEFAULT
ID	NUMBER	22	N	(null)
NAME	VARCHAR2	50	N	(null)
BRANCH	VARCHAR2	50	Y	(null)
ADDRESS1	VARCHAR2	100	Y	(null)
ADDRESS2	VARCHAR2	100	Y	(null)
ADDRESS3	VARCHAR2	100	Y	(null)
CITY	VARCHAR2	100	Y	(null)
STATE	VARCHAR2	2	Y	(null)
ZIP	VARCHAR2	10	Y	(null)
COUNTY	VARCHAR2	100	Y	(null)
WEBSITE	VARCHAR2	100	Y	(null)
PHONEAREA	NUMBER	22	Y	(null)
PHONE1	NUMBER	22	Y	(null)
PHONE2	NUMBER	22	Y	(null)
PHONEEXT	NUMBER	22	Y	(null)
DATE_CREATED	DATE	7	N	(null)
DATE_MODIFIED	DATE	7	N	(null)
MINORITY_OWNED	NUMBER	22	N	0
INDUSTRY	VARCHAR2	100	Y	NULL

In response to a query, the database returns a result set, which is just a list of rows containing the answers. The simplest query is just to return all the rows from a table, but more often, the rows are filtered in some way to return only the answer wanted.

Often, data from multiple tables is combined into one, by doing a "join". Conceptually, this is done by taking all possible combinations of rows (the "cross-product"), and then filtering out everything except the answer (Wikipedia, n.d.). In practice, relational database management systems try to be more efficient by rewriting ("optimizing") queries to achieve faster performance. The flexibility of relational databases allows programmers to write queries that were not anticipated by the database designers. As a result, relational databases can be used by multiple applications in ways the original designers did not foresee, which is especially important for databases that might be in use for decades. This has made the idea and implementation of relational databases very popular with businesses.

Views and Indexes

Complex databases use overhead data to enhance database performance, accessibility and data security. Views, as well as indexes help rearrange data in new tables for certain purposes. In an index table, related and frequently used data elements are summarized for quick retrieval. Even though views are also used to rearrange data into new tables, they are most commonly used for security purposes. By creating a view, the administrator can set viewing rights in order to restrict and manage content for different users. For example, a company might use one view containing all their sales data. The decision to create this view might be based on the frequent use of the same sales information by different sales representatives. Instead of relating data elements from different tables repetitively the sales personnel can now access one view, which creates the impression of a better database performance to the user. However, a company having one table that is accessible to all sales representatives may create a security issue. Therefore, the amount of information the sales personnel can view might be matched with their job functions and different sets of access rights might be granted. Keeping sensitive data away from unauthorized users, and hence making it more difficult for them to access, increases the security of the database.

Applications of Databases

Databases are used in many applications, spanning virtually the entire range of computer software. Databases are the preferred method of storage for large multi-user applications such as ERP systems or data warehouses, where coordination between many users is required. Even individual users find them convenient, and many electronic mail programs and personal organizers are based on standard relational database technology. As the amount of information retained and processed by organizations continues to grow, it will become increasingly important to organize the information and make it easily accessible to those who need to use it.

Technical Issues in Implementing ERP Systems

a. *Transactions and Concurrency*

In addition to showing the structure of the application by using E-R models, many databases attempt to enforce data integrity properties since system crashes and aborts do happen. In order to ensure the integrity of their data, these systems enforce the ACID rules, summarized here:

- *Atomicity*: Either all the tasks in a transaction must be performed, or none of them. The transaction must be completed, or else it must be undone (rolled back).
- *Consistency*: Every transaction must preserve the integrity constraints — the declared consistency rules — of the database. It cannot place the data in a contradictory state.
- *Isolation*: Two simultaneous transactions cannot interfere with one another. Intermediate results within a transaction cannot be visible to other transactions.
- *Durability*: Completed transactions cannot be aborted later or their results discarded. For example, they must persist through restarts of the DBMS after crashes.

In practice, however, many DBMSs allow most of these rules to be selectively relaxed to improve performance.

b. *Difficulties in Normalization*

Even though a relational database minimizes anomalies in stored data, it often suffers from performance drawbacks. The high requirements in processing power prevented its widespread adoption for some time after its introduction, although this schema has gained in popularity as technology has advanced and is now the dominant database model. Due to its complex theoretical underpinnings and a lack of knowledge on the part of the people developing databases, occasionally situations are encountered where tables are not normalized correctly and this situation can lead to major disruptions in the processing of business information.

c. *The Need for a Single Instance*

"You have ERP," says Bill McDermott, president and CEO of SAP America. "The next step is to expand it to CRM and the supply chain." The idea, he says, is to control all the data in a company by standardizing on one system for the front end and using one data source for the back end. His pitch reaches its climax when McDermott propounds his main message: "It's time to move to a single instance" (Worthen, 2003). With a single instance, all of a company's financial data will live in one application and will originate from a single source, eliminating consolidation errors and greatly reducing the time it takes to close the books. Having a single data source could also create new revenue opportunities and cut costs. Companies would be able to run reports that show cross-promotion opportunities and places where they could reuse equipment or leverage purchasing power.

There are also financial benefits; AMR estimates that companies budget an average of $4.3 million for a single-instance order management module versus $7.1 million for multiple instances (Worthen, 2003).

In the past, there were a number of reasons why a single instance was almost impossible to achieve. To start with, databases large enough to serve entire enterprises just didn't exist — at least not at prices most could afford. On top of that, connecting to that single database from faraway locations was almost impossible. In order to have a single instance of data, the same products and names have to be used throughout a company and employees need to have access to current product descriptions. Hence, along with the standardization effort, the company also needs to establish an efficient communication network. It is easy to imagine a situation where a Chinese subdivision uses the same supplier as the company's U.S. division, but neither subdivision knows this. In this scenario, it is very unlikely that both divisions would apply the same primary key to the supplier. In a single instance system, the supplier would be recognized throughout the company under a unique identifier. A single instance system reduces errors by avoiding the need to synchronize data from several databases or ERP instances.

An additional requirement is that a successful instance consolidation requires business processes and data to be simplified and standardized. The variability within business processes must be minimized and master data harmonized. As a result, successful instance consolidation requires strong corporate governance, good cross company communication, and a certain amount of business process redesign.

Once a single instance is achieved, the cost savings for the company are sizable. The IT department will need fewer people to maintain their databases, the number of licenses will be lower, and the number of locations needed to host the databases will also be reduced. They can also simplify the complicated networks that they use to consolidate the data from different sources. The company will benefit by having more time to close the books, while at the same time reducing errors. This reduction in financial errors has becomes increasingly important due to legal changes. In 2004, the Sarbanes-Oxley (2002) regulations came into effect, requiring publicly traded companies to have verifiable audit trails for their financial reports. With a single instance, all of a company's financial data will be hosted in one application and will originate from one database, eliminating consolidation errors. Because every system is integrated and automated, there is no need for employee intervention, which minimizes the opportunity for fraud. It is very likely that these types of regulations will become more prominent, making a single instance a competitive advantage for those companies that are early adaptors.

Having a single instance also makes it easier to mine customer data for trends and business opportunities. With all the data in a single place, it is easier to streamline the company's supply chain, increase the efficiency of equipment and find cross selling opportunities to mention only a few of the advantages (Worthen, 2003). In the long term, migrating to a single instance is a major investment that will be necessary for many companies to stay competitive.

d. *Integration with Non-SAP and Legacy Systems*

General purpose computers started to enter the business world in the 1950s. These big mainframe computers allowed only some simple computing functions. Since then, we have moved to multipurpose personal computers, or PCs, on every desk that are linked together in networks. Recently, the trend of untangling all the computers by using wireless connections has evolved.

As this very brief overview demonstrates, corporations have gone through some major changes within the last 50 years. One result of the rapid development and change is that corporations tried to gain a competitive advantage by developing their own proprietary systems in house. Unfortunately for them, technology has evolved so rapidly that most of their efforts quickly become obsolete compared to new off-the-shelf software. As a result, many companies that are switching to an ERP system such as SAP R/3 will still have some legacy systems that need to be incorporated. Interoperability becomes increasingly important as the IT industry is trying to make all software and hardware talk to each other.

Let us see what Thomas G. Schuessler also known as "Mr. SAP" has to say during an interview with searchsap.com:

What one technical question do clients ask most often? How do you answer this question?

Schuessler: Since we have been concentrating on integration between SAP and non-SAP components using mainly SAP's Business Application Programming Interfaces (BAPIs), most customers want to know what the appropriate architecture is for their integration project. The answer to that depends very much on the details of the customer situation: Do they have other ERP systems? Have they standardized on certain platforms and languages? SAP has many different open interfaces and using them appropriately requires a bit of analysis.

Assisting companies to integrate SAP and non-SAP components and applications is one of your specialties. Can you tell me more about your application integration issues and your solutions?

Schuessler: In most scenarios, customers will use BAPIs, maybe exclusively, or maybe in conjunction with Application Link Enabling (ALE). There are a number of services that any BAPI-enabled application needs to use. Therefore, we have concentrated on building components for Java and Visual Basic developers that offer these services nicely encapsulated and easy-to-use. This allows customers to concentrate on their applications and not have to worry too much about some of the technical details. One of the major features of our components is complete metadata support for SAP, which allows software companies to build tools around SAP easily. One of our licensees, Scribe Software Corp. (www.scribesoft.com) uses the metadata services in their Integrate tool that facilitates building bridges between SAP and other systems.

BAPIs were introduced in R/3 3.1G. For those who are new to SAP programming, please describe BAPIs and their functionality.

Schuessler: BAPI is short for Business Application Programming Interface. A BAPI is a method of an SAP object type that can be called from an external program. An example would be SalesOrder.CreateFromData which enables you to build your own front-end for order taking, for example in a Web application running on a Java application server (Brunskill, 2001).

e. *Complexities Due to the Widespread Use of E-Commerce*

The above interview mentions e-commerce, which refers to business that is conducted over the Internet. Amazon is one of the best known examples of a company that does its business solely via the Internet. A reduction in the corporate cost structure makes it very attractive to conduct business online. The big advantage of doing business via the Internet is that every transaction is recorded in digital format, allowing it to interact directly with the ERP systems. Many companies such as airlines provide incentives (such as airline miles and cost advantages) to encourage end-users to use the Internet to make their travel arrangements instead of calling their telephone agents.

f. *The Need to Mine Data Warehouses*

One of the major benefits that an ERP system provides is that all the company information is in digital form and can be available instantaneously. Even though this has risks, such as security and reduced fault tolerance, it also allows technologies such as e-commerce and/or data warehouses to be integrated with the ERP system. Once all the transactional company data are captured, it is logical to store them and use business intelligence tools to find patterns that can help improve the business. Highly sophisticated software will crawl through the terabytes of data to find patterns that can be of value to employees. The Internet has made it possible to distribute the information gathered from the data warehouse to anybody, anywhere in the organization. Although this has the potential to drastically increase the efficiency of every worker, this also increases the potential security risks.

g. *Increasing Focus on Security Issues*

Allowing users to access the data warehouse via the Internet means that security has become even more critical. Security has two main aspects. First, the data should be safe from malicious parties. Malicious parties not only includes people attempting to break into the data warehouse, but also clients or other parties either accidentally or willfully attempting to gain access to data that they are not meant to see. The second aspect of security deals with preventing accidental data losses to ensure the data's consistency and quality and its safe conveyance to those who need to use it.

Common Security Concerns

The most obvious security concern is securing the data. It has to be safe from unauthorized users and access should be restricted to only the users or groups of users who require the data. Next is securing the network. Once the data leaves storage it is open to attack and requires much more effort to secure than the data that is not being used. Users have to be classified and assigned privileges based on their needs.

Physical vs. Virtual Security:

The physical security of data and networks include restricting access to the hardware through safety features such as biometric scanners or ID cards. It is also important to secure the IT assets from external environmental impacts such as floods, heat, power outages, and power spikes. Once the organization has ensured that the datacenter is safe from physical intruders and external forces, the IT architecture needs virtual protection to prevent intruders from accessing these assets via the Internet. Some of the tools used are proxy servers, firewalls, routers and honey pots. Additional security measurements need to be taken if the organization operates a wireless network. Wireless security should include access logs, encryption and directional antennas.

Social Engineering

Most organizations are aware of the security threats discussed above, but forget to include their most vulnerable assets, namely their employees. Intruders do not necessarily need great technical skill to extract data from a company or cause harm to its IT assets. A common technique used when hacking into an organization is the use of social engineering. This approach uses a combination of social skills and inside information to break into the organization and extract information or harm the assets. A famous hacker who is well known for his social engineering skill is Kevin Mitnick. He is the author of *The Art of Intrusion* (2005), where he describes the basics of social engineering. It is very important to make employees aware of this threat to prevent companies from suffering major data losses.

Performance vs. Security

In any data warehouse, the main emphasis is on the two most important aspects — security and performance. While implementing a data warehouse, security is very important and should be given due consideration in the infrastructure, but the fact remains that the main point of building a data warehouse is to provide easy and efficient access to its users. As a result, it is not wise to implement security at a level which compromises performance. On the other hand, there is no point in providing greater performance at the cost of exposing the data warehouse to unnecessary security risks. So, the implementation of a data warehouse is a delicate balance between the performance required and the security necessary. Theoretically speaking, a perfect security model is very simple. It should not allow any unauthorized user to access data; data should be secure, and any attempts to breach the security of the data warehouse should fail without exception. Authorized users should feel confident that the data they are receiving is free of errors, accurate and complete. Unfortunately, security threats will always be present. The key is finding an acceptable balance between optimum performance and the level of security needed to ensure the data is appropriately protected.

Implementation of SAP R/3

As we discussed earlier, the implementation phase is very important and one of the key success factors. Thus, SAP has defined a roadmap for the implementation of the R/3 system which is called ASAP (Accelerated SAP). The ASAP was first introduced in 1997 as SAP's recommended implementation strategy. It is a phased implementation strategy that integrates project knowledge and practical experience and uses tools and services for rapid implementation and continual optimization of R/3 installations.

ASAP distributes responsibility for the implementation among three people. The first is the project manager, who is responsible for the planning and execution functions. Second, an application consultant is needed to deal with the creation of the business blueprint, the customization and configuration of the R/3 system, organizing team member training and running executive meetings. The application consultant works with the project team. The last role is the technical project leader, who is responsible for technical issues such as the definition of the system prerequisites, configuration of the system landscape, and technical training of the system administrators. This role distribution reduces redundancy, helps to achieve targets more efficiently and ensures the synchronization of all activities (Buck-Emden, 2000, pp. 244-252).

There are five phases in ASAP (Figure 4): Project Preparation, Business Blueprint, Realization, Final Preparation, and Go Live & Support. There are also ASAP tools that are used throughout the implementation process that help to control the phases. Project Estimator is used by SAP consultants to monitor the resources, costs and time. This tool is mainly used before the project starts to estimate the project costs. Implementation Assistant helps to control every step and ensures that it is performed correctly. R/3 Configuration Tools manage the configuration process and are used in the realization phase.

Figure 4. SAP's implementation roadmap (SAP AG, 2005)

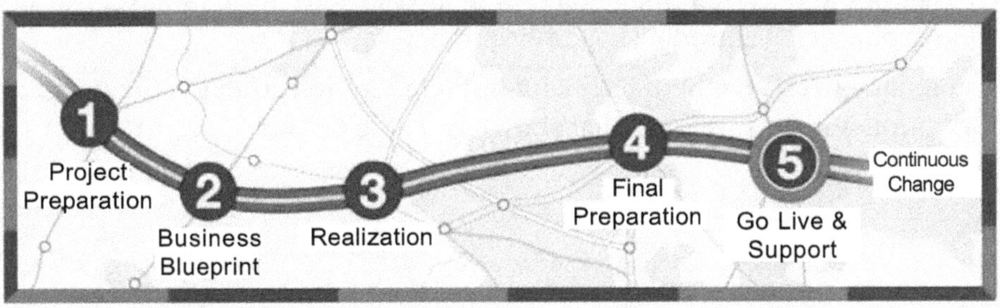

1. *Project Preparation*

In this phase, the scope of the project is defined and clear project objectives are set. The organization standards for the project are defined and the necessary technical requirements are determined. In this phase, proper planning and organizational readiness assessment are essential. This phase also identifies and plans the focus areas that need to be taken into consideration.

2. *Business Blueprint*

The purpose of this phase is to create a common understanding of how the company intends to use SAP for its business. The business blueprint does this and documents the company's business process requirements. Business blueprints describe how the implementation will take place and how it will support specific business processes. The end result is a comprehensive blueprint of the business. In this phase, the project objectives defined in the first phase are finalized.

3. *Realization*

The business process requirements that were defined using blueprints in Phase 2 are implemented in the realization phase. The aim is to finalize the configuration of the R/3 system by setting up all the parameters and defining the basic data. The system will also be tested and released in this phase.

4. *Final Preparation*

The purpose of this phase is to finalize the organization's readiness to "go live" and finish all the preparations. This includes testing the system, training the users, and systems management. All critical open issues must be resolved in this phase. With the successful completion of this phase, the system is ready to go live.

5. *Go Live & Support*

With the beginning of this phase, the project oriented approach ends and all necessary improvements in the production system are complete. Issues such as user training and system management also continue in this phase, but unlike previous phases, in a production environment.

If a company uses a 6 month big-bang implementation strategy, the typical time distribution will be similar to that shown in Table 6.

Table 6. Time distribution for a six month long big bang implementation (Buck-Emden, 2000)

Phase	*Duration*	*Percentage of overall project time*
Project Preparation	1 week	5%
Business Blueprint	4 weeks	15%
Realization	16 weeks	61%
Final Preparation	5 weeks	19%
Total:		100%
Go Live & Support	Approx. 2 weeks	

Enterprise Modeling as a Part of the Blueprint Phase: Structuring of a Company

The R/3 system applications consist of business processes, organizational units and information. Business processes define the steps needed for each certain process, such as the make-to-order process. Organizational units control business processes and information flow. Information defines the amount of data that must be stored, retrieved, and processed in each application.

Figure 5 illustrates the allocation and shows that in each module the importance of these elements differs.

Figure 5. The allocation of business processes, organizational units and information according to modules (SAP AG, 2005)

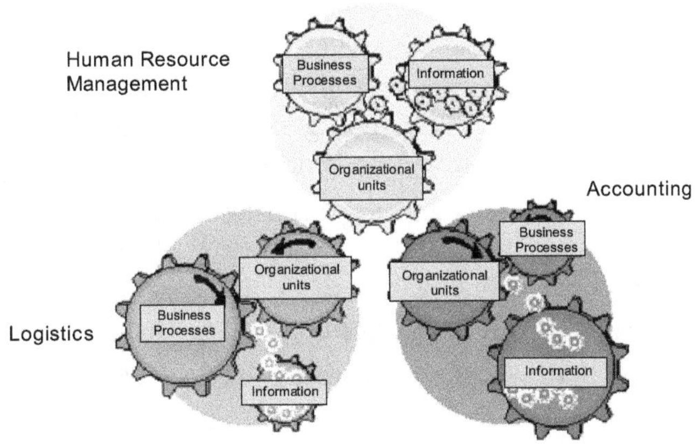

Table 7. Mapping a company to organizational units of R/3 (SAP AG, 2005)

Business Structure Elements	R/3 organizational units
Enterprise	The R/3 System
Company	Operating concern Controlling area Company code
Business unit	Business area
Business location	Plant

The importance of business processes is greater in logistics, whereas information flow is very important in accounting for internal and external reporting requirements. All the elements in the HR module are equally important.

Before the implementation can proceed, the structure of the company must be simulated with organizational units so that business processes can run smoothly and efficiently. This also ensures that the internal and external reporting requirements are met. Table 7 is an example of such mapping.

The R/3 system aligns with the whole company and all functions in the company are mirrored in R/3 modules. Profitability issues are handled under operating concern, whereas cost center issues are handled under controlling area. A company code is the smallest organizational unit for which a complete self-contained set of accounts can be drawn up for purposes of external reporting. Internal balance sheets for the business units can be created by business area. A location where products or services are provided can be represented with a plant.

Another important concept that is relevant for the organizational units is the Strategic Business Unit, which by definition is the smallest unit in a company that has its own market and capital. The mapping of organizational units must follow the structure imposed by the strategic business units.

Company codes, business areas and profit/cost center are accounting related organizational units. Logistics related organizational units are plants, sales organizations and purchase organizations.

A fully represented example is given in Figure 6. This example is from SAP's documentation, which illustrates a business area scenario in a simple and effective way.

Figure 6. Business area scenario example (SAP AG, 2005)

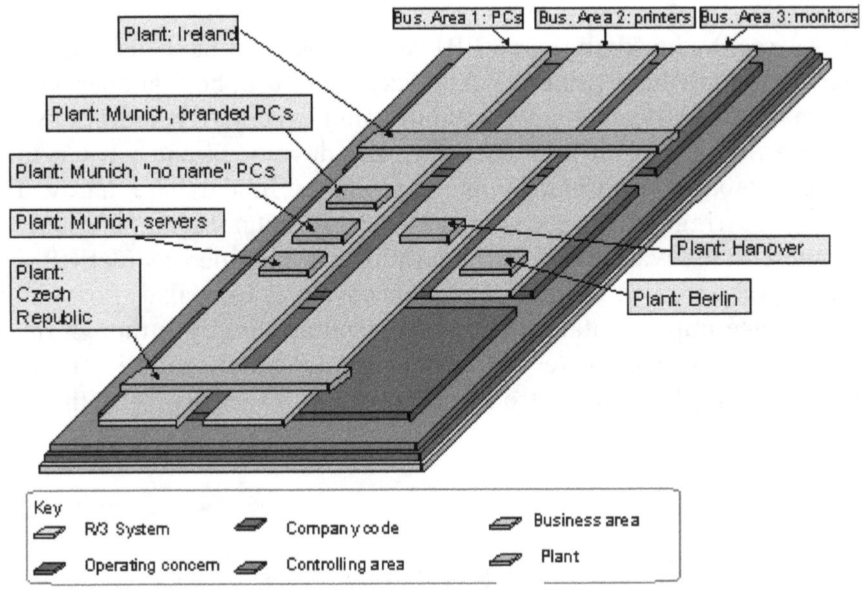

The company shown in Figure 6 is represented with the R/3 system as the organizational unit at the very top. The controlling area includes every company code, business area and plant. This company has three different company codes representing its three legally independent companies located in Ireland, Germany and the Czech Republic. The company has three business areas that manufacture PCs, printers and monitors. On top of these organizational units, plants are defined. For instance, business area PCs have three different plants in company code Germany meaning the company has three companies in Germany producing or selling PCs. The Czech Republic doesn't have a business area for monitors, meaning that the company does not produce monitors there. In Ireland, there is only one plant that produces all three products.

This example shows the importance of selecting the right system units to represent a particular company. The structure depends on the business strategies that the company uses. This emphasizes that these structural decisions have to be made in the blueprint phase, and if the new system is implemented using the wrong structure, the company will not be able to change it easily.

Chapter Summary

Databases are the backbone of an ERP system. As you have been reading this chapter, you should have gained a good understanding of how databases organize data to make it easily accessible without using excessive resources. You also followed a discussion on how databases are developed and what it takes to optimize database performance. New advances in database hardware and software have triggered the convergence of many technologies, and some of the most important innovations are the integration of e-commerce with ERP systems and data warehouses. In addition, technical advances have also allowed for true single instance implementations in large corporations. Even though the latest technological advances have contributed tremendously to the efficiency of corporations, they have also introduced a major new security threat that must be addressed by companies.

Team Assignments and Discussion Questions

1. Discuss the need for a DBMS.
2. Define and discuss the use of a relational database.
3. Discuss normalization. Provide an example.
4. What is the difference between a logical and a physical model?
5. Define and provide an example of the E-R model.
6. List and discuss three technical issues involved in implementing ERP systems.
7. Describe the implementation roadmap that SAP uses for implementations.
8. Provide an example of structuring an organization using SAP R/3's approach.

References

Asia Pacific Women's Information Network (APWINC) (2002). Data process system. Retrieved September 7, 2005, from http://www.apolc.org/computer/database/main1-2.html

Brunskill, V. (2001, May 8). Inside SAP. An exclusive interview with Thomas G. Schuessler. Retrieved from http://searchsap.techtarget.com/qna/0,289202,sid21_gci550189

Buck-Emden, R. (2000). *The SAP R/3 system: An introduction to ERP and business software technology* (pp. 244-252). London: Addison Wesley.

Carkenord, B. A. (2002). *Why build a logical data model.* San Francisco: Embarcadero Technologies, Inc. Retrieved September 7, 2005, from http://www.dmreview.com/whitepaper/dmd.pdf

Chen, P. P. (1976). The entity-relationship model – Towards a unified view of the data. *ACM Transactions on Database Systems*, *1*(1), 9-36.

Fagin, R. (1981). A normal form for relational databases that is based on domains and keys. *ACM Transactions on Database Systems*, *6*(3), 387-415.

French, W. L., Bell, C. H. & Zawacki, R. A. (2000). *Organizational development and transformation: Managing effective change* (5th ed.). New York: McGraw-Hill.

McDonough, A. M (1969). *Centralized systems: Planning and control.* Wayne, PA: Thompson Book Company.

Mitnick, K. D. & Simon, W. L. (2005). *The art of intrusion: The real stories behind the exploits of hackers, intruders and deceivers.* Indianapolis, IN: Wiley.

Relational Database (2004, September 27). Relational database technology: A crash course. Retrieved September 7, 2005, from http://examples.oreilly.com/pnetwinaps/appa.pdf

Sankar, C. S. (1985). Analysis of names and relationships among data elements. *Management Science*, *31*(7), 888-899.

Sankar, C.S. (1991). Evaluation criteria for data dictionaries. *Journal of Database Administration*, *2*(1), 1-6.

SAP AG (2005). Enterprise modeling – consultant's handbook. *SAP documentation.* Retrieved July 24, 2005, from http://help.sap.com/saphelp_46c/helpdata/en/7e/d4342045591 1d189400000e8323c4f/frameset.htm

Sarbanes-Oxley Act (2002). U.S. bill HR3763. Retrieved September 5, 2005, from http://news.findlaw.com/hdocs/docs/gwbush/sarbanesoxley072302.pdf

Singh, H. (1998). *Interactive data warehousing* (p. 163). Upper Saddle River, NJ: Prentice-Hall.

Wikipedia (n.d.). Retrieved September 7, 2005, from http://en.wikipedia.org

Worthen, B. (2003, November 15). Extreme ERP makeover. *CIO Magazine.* Retrieved September 7, 2005, from http://www.cio.com/archive/111503/erp.html

Appendix A

Research on Data Element Definition in Academics

Faculty members in different universities have long had strong opinions about the use of computers in industry, and the attitude of some faculty during the 1970s was that the computers would be used to invade the privacy of individuals, automate companies, and make life dull and boring. At the same time, a few faculty members who had military backgrounds or who had worked in industry believed that it was possible to centralize the information systems and standardize the technical terms that were used in companies. A pioneer in that effort was Adrian M. McDonough, a professor of management at the Wharton School of Business of the University of Pennsylvania. In his book *Centralized Systems: Planning and Control* (1969), he devised a management search pattern to solve "information inventory" problems. He suggested a methodology to develop a corporate memory that documents the contents of the information systems in an organization. Based on the methodology developed in this book, one of the authors of this book, Chetan Sankar, developed a syntax methodology that can be used to resolve any conflicts in defining elements and the relationships among elements. The Office of Management and Budget in the Executive Office of the President of the United States was interested in the syntax method and provided data entry forms that were used by federal agencies for this research project. The outputs of the method become the consistently defined data elements (Sankar, 1985).

Evaluation Criteria for Data Dictionaries

There are no established criteria to evaluate how well data dictionaries and repositories eliminate redundancy and help retrieve meaningful data elements in response to a query. This paper proposes five criteria — namely, relevance, consistency, common use among information systems, degree of automation, and degree of security — to evaluate data dictionaries and repositories, and explains them using a detailed example. Empirical studies that apply these criteria to data dictionaries used in an organization would further refine the criteria and provide valuable results to data base administrators.

Proposed Criteria to Evaluate Data Dictionaries/Repositories

The proposed criteria measure how well a dictionary/repository collects, maintains, and retrieves information about data. Its objectives are to offer users cost-effective means of retrieving relevant information and reducing information overload. Based on these objectives, four criteria are proposed to evaluate data dictionaries/repositories: relevance, consistency, common use among information systems (IS), and degree of security. Maximum size of elements, hardware requirements, and other limitations of the products are not proposed as criteria, since this information is easily obtained from the vendors. Instead, a DBA will obtain valuable information if historical data were collected about the products using the five criteria.

Each criterion is defined and its applications elucidated by the example that follows. Consider a mail order company whose order, catalog, and billing departments maintain different data dictionaries. These departments store and retrieve information about a customer for their individual operations. The order department requests information about the customer's name, address, sex, order information, and method of payment when a customer calls in an order. The catalog department mails catalogs to customers on the basis of their age, salary, family status, amount of annual order, special tastes, and other information. The billing department uses information about the customer's frequency and promptness of payment by credit cards, checks, or money orders to establish their credit history. Top management wants to substitute a central data dictionary/repository for these three data dictionaries. They assign this task to a DBA, who has now to choose an appropriate repository/dictionary that could serve not only these three but others in the future. The DBA short-lists two products as possible candidates for company-wide use and compares their performance by standardizing and storing elements about customers in the three departments. The DBA uses these two products to combine the elements and makes the resulting data dictionaries available to the users. Statistics about the performance of the products based on the following five criteria are collected. The following criteria measure how well each commercial repository/dictionary product fulfills the DBA objectives.

a. *Relevance*: This criterion measures the effectiveness of retrieving correct information in response to a request. Relevance is measured by two factors: recall and precision. These factors are popularly used in the library science area and have been modified here to suit the environment of data dictionaries/repositories.

$$\text{Recall} = \frac{\text{Number of matching data elements retrieved by a product}}{\text{Maximum number of matches possible}}$$

$$\text{Precision} = \frac{\text{Number of matching data elements retrieved by a product}}{\text{Number of data elements retrieved by a product}}$$

These criteria will be explained with a possible query to the data dictionaries created using the two products. Suppose the company is test-marketing a new English-German hand-held electronic translator. They decide to offer this product to a selected set of customers. They request the DBA to provide a list of white collar customers who are older than 30, travel outside the country, speak either English or German but not both, are married, pay their bills regularly, and have annual income above $30,000.

A staff of the DBA manually identifies four matching data elements for this query: married, payment record, salary, and age. The other elements are not stored in the data dictionaries. When the staff use products A and B to identify the number of matching data elements, the results are not the same. Assume the following results are obtained.

Maximum number of matches possible = 4

Product A:

Number of data elements retrieved = 6; Number of matching data elements = 3
Recall = 3/4 = 75%; Precision = 3/6 = 50%

Product B:

Number of data elements retrieved = 60; Number of matching data elements = 4
Recall = 4/4 = 100%; Precision = 4/60 = 6.7%

Though product A retrieves 6 elements, only three of them (married, salary, and age) match with the maximum possible matches. Product B has better recall since all the possible matches are retrieved, but at the cost of very low precision. Therefore, the DBA may look for other criterion to compare the products.

b. *Consistency*: This criterion measures the ability of the standardization products to identify and store all primitive data and public and recurring derived data, and no more. Primitive data describe single entities, whereas derived data describe multiple entities. A primitive data element directly depends on a single occurrence of the entity or an event, such as information about each order, and particular details about a customer (age, sex, and salary). Derived data depends on multiple occurrences of an entity; for example, a customer's credit history, payment balance, and average value of purchases during the past 10 years are derived data. Although credit history and payment balance are derived data, they are stored as primitive data in the company's public domain so any employee can quickly retrieve them. But, because the probability of retrieving a customer's average value of purchases is low, that information need not be stored in the data dictionary. If needed, it can be computed from the primitive data. Primitive and derived elements are also called as elementary and optional elements.

Even among the primitive data elements, duplicate elements need to be identified and removed; confusions about the meaning of primitive elements need to be resolved. For example, expenses, expenditure, and amount spent may all be stored as primitive elements; examination of these elements may indicate that they are duplicates and can be represented by one of these three names. Confusion about the meaning of primitive elements can happen when a customer is coded in a dictionary using the social-security number, whereas in another data dictionary is represented using an account number of that customer. A unique identification scheme needs to be established for each customer and needs to be consistently used in all the data dictionaries. The result will be a canonical data dictionary — a minimal and nonredundant representation of data elements in an enterprise.

Consistency measures the performance of the products in removing redundancies and storing the minimum number of elements from which all other elements can be derived.

$$\text{Consistency} = 1 - \frac{\text{Number of elements in the final data dictionary}}{\text{Number of elements in the original data dictionary}}$$

The dictionary in the mail order example could be made consistent by storing only the nonduplicate primitive and public recurring derived elements. Assume that a member of the DBA's staff manually classifies the elements, and finds thirteen primitive and public recurring derived data elements: customer's name, address, sex, mode of payment, order date, order item, order quantity, order price, date of birth, salary, wages, income, and family status. Of these salary, wages, and income are duplicates and are replaced by one primitive data element — salary.

Though order price can be derived from the cost of the product, profit margin, and overhead costs, it is a recurring public data and hence is stored as a primitive data element. Thus, 11 primitive and derived recurring data elements need to be stored. All other data elements are derived or duplicates. Examples of derived elements are: value of annual order, special tastes, age, and frequency of use of credit cards, checks, or money orders. The value of annual order is derived by averaging the value of each order. The special tastes can be identified by analyzing the item purchased and quantity of purchase. Age can be computed from the primitive data element, date of birth. Frequency of use of credit cards, checks, or money orders could be computed from the mode of payment of each order.

The DBA evaluates the two data repository/dictionary products by comparing how the products had standardized the data dictionaries.

Number of elements in the original three data dictionaries = 120

Product A:

Number of elements in the final data dictionary = 10;

Consistency = 1 - 10/120 = 91.66%

Product B:

Number of elements in the final data dictionary = 70;

Consistency = 1 - 70/120 = 41.66%

Note that Product A failed to capture one primitive element of the 11 identified by the DBA's staff, and, product B stores many more elements in the dictionary. Another important criterion is whether the products can be used to standardize multiple information systems (IS).

c. *Common use among different IS*: This criterion measures whether the product can be consistently applied to standardize IS in different departments and operations within an organization. Current trends toward integrating networks and information systems to build integrated repository-network management environment makes it important that data dictionaries and repositories handle multiple environments. Some data repositories/dictionaries are tailored to particular operations and may not be usable in other areas. Hardware and software products sometimes incorporate prepackaged dictionaries. For example, microcomputer-based word processing software, spread sheets, and database systems often have built-in data dictionaries. Elements defined with these dictionaries cannot be easily transferred to organization-wide dictionaries/repositories. Deciding which dictionary to use as the central repository may depend on a dictionary's flexibility in handling a variety of software and hardware. The common use criterion measures this flexibility.

$$\text{Common Use} = \frac{\text{Number of elements standardized using a particular product}}{\text{Number of elements standardized in the organization}}$$

Assume that the mail order company has expanded its business to international locations. The DBA has been asked to extend the implementation of the two data dictionary products to five countries and to store information about the performance of the telecommunication networks. Management wants answers to such questions as: Which country produces the most orders, which carrier offers the cheapest long distance rates, and which telecommunications facilities fail often? The DBA uses both products to standardize information about the countries. Assume that the number of primitive and public recurring derived elements among all the relevant data dictionaries are now 1,500, but that each product can standardize only a portion of these elements.

Product A:

Number of elements standardized by Product A = 1,000;

Common Use = 1000/1500 = 66.7%

Product B:

Number of elements standardized by Product B = 400;

Common Use = 400/1500 = 26.7%

Product A is designed to work across different domains; hence the high percentage on common use. Product B has a lower percentage since it is specially designed to standardize elements in business use and cannot document technical terms used in telecommunications. Constraints on budget and manpower may favor products that are automated, and the next criterion measures the degree of automation of data dictionaries.

d. *Degree of Security*: Illegal access to data elements is a serious problem for an organization. Competitors can gain information about the data structure and data model of a company once they access the data repository/ dictionary. Employees of the company also should be restricted in accessing the primitive elements and in deriving other data elements. Overall security depends on managing the access controls to the various primitive and public recurring derived data elements. Access control limits have to be defined for each user and violations acted upon. It is necessary to do experiments that simulate the penetration attempt through the access control limits. On the basis of such experiments, the following criterion is proposed.

$$\text{Degree of Security} = 1 - \frac{\text{Number of security failures}}{\text{Number of attempts to breach security}}$$

Assume that a member of the DBA's staff tries to breach security 150 times.

Product A:

Since product A's data dictionary is intended to be used in many systems, security checks are built into the software.

Number of security failures = 2; Degree of Security = 1 – 2/150 = 98.67%

Product B:

Since product B's data dictionary is a part of an application software, no special efforts have been made to secure the dictionary from any authorized user of the software. The DBA's staff have written programs to improve the security of the data dictionary.

Number of security failures = 30; Degree of Security = 1 – 30/150 = 80%

Comparing both products with the data obtained for each criterion can provide the DBA with relevant information for choosing a data dictionary.

Discussion

The performance of the two data dictionary products on each criterion is summarized in Table 1 of Appendix.

A DBA can use these criteria to select the appropriate product for the application at hand. For an application, the DBA might generate a list of expected values for

Table 1.

Criterion	Product A	Product B
Recall	75%	100%
Precision	50%	6.7%
Consistency	91.66%	41.66%
Common use	66.7%	26.7%
Degree of Security	98.67%	80%

each criterion, match those expectations with the performance of the products, and choose the product with the values closest to the expectations. Assume, for example, that a DBA must choose a company-wide repository. The DBA estimates that for such an application, the product should have reasonable recall (70%), high precision (80%), high consistency (90%), high amount of common use (90%), and excellent degree of security (98%).

These expectations were compared with the table showing the performance of the two products in order to choose the appropriate product. Product B may not be a serious candidate due to this product's lower values for precision, consistency, common use, and degree of security. The higher scores on recall of product B may not sufficiently compensate for the very low scores on the other criteria. Product A might be a reasonable choice since it surpasses the DBAs expectations in recall, consistency, and degree of security. This product does not meet the DBAs expectations in precision and common use. Given these results, the DBA might expand the list of choices and include another product as a candidate. Another alternative is for the DBA to make more realistic estimates of expectations.

Historical statistics maintained on these criteria for each data dictionary in use in an organization could be helpful to the DBA in choosing among the products. Collecting such statistics also would help new members of the DBAs staff in choosing appropriate data dictionaries/repositories for use in the organization.

Chapter VI

Multiple Information Systems for Coping with a Growing and Changing Business: Robert Bosch GmbH

Learning Outcomes

- Understand the importance of aligning the business and IT strategies in a company at both corporate level and country level
- Identify the issues faced by RBUS in moving to a standardized corporate-wide information system such as SAP R/3 during 1992-1999

Case Overview

In early 1998, Don Chauncey, the vice president and chief information officer (CIO) for Robert Bosch Corporation (RBUS), faced a serious problem. He had just talked with Computron Software, Inc. (http://www.computronsoftware.com) executives, who had told him that they were going to support their financial systems software on IBM AS/400 computers only through December 2001. In the future, Computron wanted to provide these functions on a client-server platform.

Don was also under pressure from the RBUS Human Resource (HR) function, who were requesting applications that were not possible using the current systems, which were operated on an outsourced IBM mainframe computer. The HR and IS staff were asking to be allowed to update to PeopleSoft in order to address the company's payroll and human resource needs.

Krish Kumar, the director of Corporate Information Technology (CIT) for the Automotive Original Equipment Division (UO), was also evaluating the company's information systems in order to address customer needs. The current nonintegrated system solutions had created pockets of implementation gaps between the business entities. Not only had they created complexity in the organization, but the old reporting methods also added time to the production of performance measures. Furthermore, the trend for Original Equipment Manufacturers (OEMs) was to have their customers deliver total systems and not just parts. Don and Krish had to establish a strategy to make the multiple IS demands work together in order to cope with the growing and changing business.

Case Study

Introduction

The name Robert Bosch is closely associated with the automobile. By 1999, however, Bosch had become known not only for automotive equipment, such as anti-lock braking systems (ABS), brakes, fuel-injection technology, and driver information systems, but also for a whole range of other product areas, such as communications technology, power tools, household appliances, thermotechnology, automation technology, and packaging machinery.

The international orientation of the Bosch Group is in line with a tradition that is almost as old as the company itself. Even before the turn of the previous century, Bosch had already established its first foreign offices, and it began construction of its first factory in the United States in 1909. In 1914, the capacity of this plant exceeded that of the German Bosch plants. In 1999, Bosch was composed of approximately 250 subsidiaries and affiliated companies in 48 countries. The Bosch Group has 185 production plants worldwide, of which 142 are located outside Germany and are spread across Europe, North and South America, Africa, Asia, and Australia. Throughout the world, Bosch holds an interest in an additional 37 joint-venture companies. An essential constituent of Bosch corporate policy is the maintenance of its innovation capability — based primarily on its large staff of experienced employees. Around the world, approximately 15,700 Bosch scientists, engineers, and technicians are working to improve and enhance the functionality and reliability of existing products, as well as on the development of new products and systems.

Business Sectors

The great number and variety of activities in the Bosch Group fall into four business sectors (UB - Unternehmensbereich). Each business sector is divided into a number of divisions/business units (GB - Geschaeftsbereich).

Automotive Equipment Business Sector (UBK)

Automotive equipment forms the largest business sector in the Bosch Group. In 1998, this sector employed 123,000 associates and achieved 63% of total sales for the Bosch Group. Bosch is one of the world's largest manufacturers of automotive equipment. The divisions (GB) in this section are:

- ABS and braking systems (K1)
- Lighting technology (K2)
- Engine management for gasoline engines (K3)
- Bodywork electrics (K4)
- Fuel-injection technology — diesel (K5)
- On-board electronics (K6)
- Mobile communications (K7)
- Semiconductors and control units (K8)
- Starters and alternators (K9)
- Aftermarket products, test equipment and after-sales service (KH)

Each of these divisions is composed of a number of product groups (Bosch GmbH, 1998).

Communication Technology Business Sector (UBC)

Bosch employs 18,700 people in the communication technology business sector, which in 1998 accounted for 5 billion DM (Figure 2) of total Bosch Group sales and investment in research and development, totaling 600 million DM. The divisions in this sector are:

- Private networks
- Public networks
- Broadband networks
- Terminals
- Aerospace technology
- Security systems

Consumer Goods Business Sector (UBG)

The share of total sales attained by the consumer goods business sector was 22% in 1998. In the field of power tools, Bosch has achieved a leading position in

Figure 1. Sales UBK in billion DM

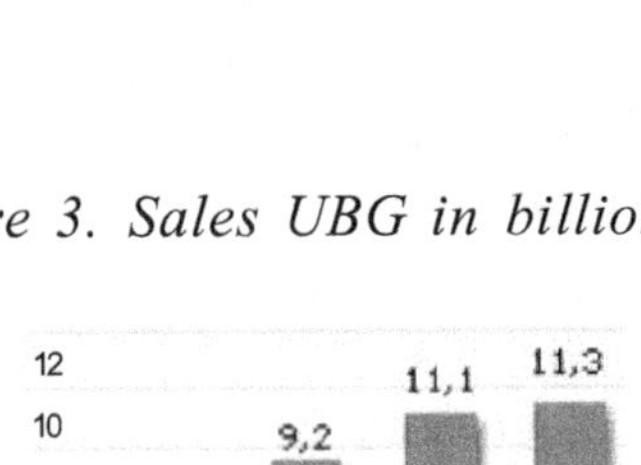

Figure 2. Sales UBC in billion DM

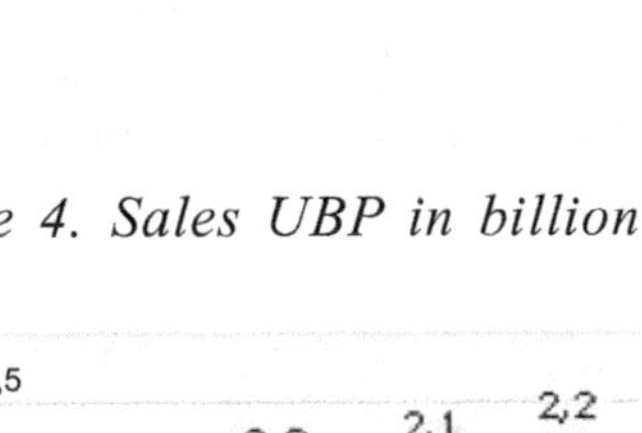

Figure 3. Sales UBG in billion DM

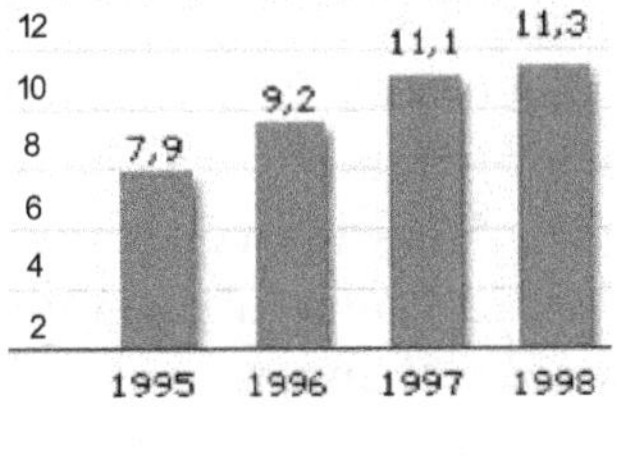

Figure 4. Sales UBP in billion DM

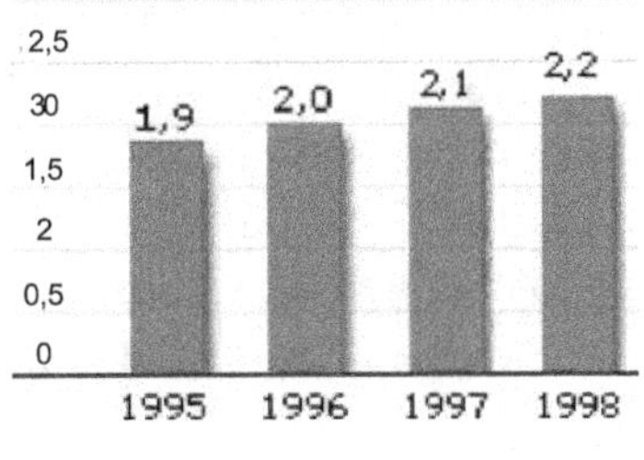

international markets. In thermotechnology, Bosch chose to concentrate on gas as the energy source for heating and hot water, with outstanding environmental compatibility. The company Bosch-Siemens Hausgeräte GmbH (BSH) is responsible for household appliances, with Bosch and Siemens holding equal interest in the business. The divisions in this sector are:

- *Power Tools*: Electric power tools for the trades and gardening, and do-it-yourself markets; tool attachments, and accessories
- *Bosch Thermotechnology*: Heating and hot water equipment, closed- and open-loop controls, gas controls
- *Household Appliances*: Refrigerators and freezers, washers and dryers, stove tops, dishwashers, small appliances

Capital Goods Business Sector (UBP)

In 1998, the capital goods business sector achieved sales of 2.2 billion DM. The Bosch Packaging Machinery Division, in particular, achieved a leading international position with its highly specialized systems. The divisions in this sector are:

- *Automation Technology*: Automotive hydraulics, industrial hydraulics, pneumatic systems, assembly engineering, drive and control engineering, deburring systems, tightening and press-fit systems
- *Packaging Machinery*: Packaging machines and equipment; machinery for the packaging of food products and pharmaceuticals

Top Management and Its Directive to the IT Division

Dr. Hermann Scholl, F1, leads Robert Bosch GmbH (RB). He is the chairman of the Board of Management (Geschaeftsfuehrung). The other board members are shown in the organization chart in Figure 5. Each board member is responsible for specific divisions, multiple countries (called RG — Regionalgesellschaft — national companies), and some of the corporate departments.

Facing problems with legacy systems in RB, the Board of Management gave two mission statements to the IT management team during 1994. First, the IT division had to operate like an external IT provider and meet or exceed benchmarks of these providers. Second, it had to be a leader in setting policies, procedures, and standards for IT throughout the Bosch Group. The Board gave the division a chance to change its own environment and establish its own future, rather than have an external consultant come in and recommend changes.

Figure 5. General organization chart of Robert Bosch GmbH in 1998 (Bosch internal document)

Group Executive Level	Corporate Responsibilities	Divisions	Subsidiaries Abroad
F1 Scholl	Chairman Coordination UBK Coordination Original Equipment Sales Executive Personnel (LD)	Semiconductors & Electronic Control Units Original Equipment Sales K8, TV	RG/TV
F2 Todenhofer	Human Resources & Social Welfare Legal Affairs & Taxes Public Relations		
F3 Hoffmann	Business Administration Finance & Finacial Statements Planning & Controlling Purchasing & Logistics Coordination W for UBK	Lighting Technology K2	
F4B Zimmerer	Production Coordination Capital Expenditure Environnmental Protection	Management Systems for Gasoline Engines Diesel Fuel-Injection Equipment K3, K5	MICO, RBIN
F4C Oswald		ABS & Braking Systems K1	RBME, RBUS
F4E Rojahn		Bodywork Electrics Body Electronics Mobile Communications Thermotechnik K4, K6, K7, TT	RBAU
F4F Dais	Research & Development Advance Engineering Coordination Engineering K Information Technology	Starting Motors & Alternators Communications Technology K9, UC	
F5A Hahn	Construction & Buildings Quality Assurance Coordination T for UBG & UBP	Electric Power Tools Automation Technology Packaging Machinery EW, AT, VM	RBBR
F5C Romberg	Coordination V (Sales) & W for UBG & UBP Sales Organization	Aftermarket Sales KH	RBFR, RBGB, RBIT, RBOS, RBSP, other V-GR, GR/H
Manger			RBJP, RBKR, VBP, VBS, RBCT, RBHK

IT Management Team at Robert Bosch GmbH, QI

The IT management team created the QI (Querschnittsbereich Infomationsverarbeitung — Division for Information Technology) business unit in order to address these requirements.

There are three units in QI. Dr. Eggensperger is the chairman of the QI executive management team and head of the unit in charge of staff functions, computing centers and infrastructure, and customer services. Mrs. Bauer takes care of the application systems unit, applications systems division, and commercial functions division. Mr. Mierzwa is responsible for the business support unit with controlling, accounting, and IT-purchasing. There are IT departments in every location that help internal customers. They have to know the processes used by their customers and be able to add value by telling them what technology and systems can be used to improve operations. They are effectively undergoing a transition from being IT technology driven experts to becoming communicative consultants. Mrs. Bauer discusses the need for change in the system:

In January 1994, we started looking for a new ERP (Enterprise Resource Planning) system. We were still happy with our existing ERP system, but we recognized some problems. The different locations used their own systems and had optimized them quite well. They had their own little paradises. But we had pressure from the Board of Management to report by certain dates, and pressure from the market for them to work together. That required us to establish interfaces that made the old system so complex that it was impossible to maintain. Some of the needs coming from the market today cannot be managed with these old systems. If you need to determine what is available in the network, you cannot determine where it might be. If someone in the network needs a product for a special project, the system will not help. If you want to establish e-commerce, it won't work. If we want to purchase new companies, we need to be more flexible. Too much time was required to make the old system work, so we needed a more flexible system. Most important, we needed an application architecture that starts from the top and goes to the bottom.

In 1994, we felt the internal pressure from the IT people who were trying to make the old systems work. In the past, we always evaluated the present situation and determined if the IT system was working. Now, we are also evaluating whether these systems will work in the future. We knew that the coordination was not working. Since it takes time to develop these systems, we have to look into the future. When we looked at the need in 4 or 5 years, we realized that the old systems were not going to work.

Structure of the Robert Bosch US (RBUS)

The chairman and CEO of RBUS is Robert S. Oswald, denoted F4C in Figure 1. RBUS has a total of 55 manufacturing plants, corporate and sales offices, technical centers, and distribution centers in 18 U.S. states, Canada, and Mexico. In 1998, it employed 23,400 employees and had sales of $5.3 billion. Of these sales, Robert Bosch Corporation generated $3.6 billion as a legal entity, and the rest was generated from operations in Mexico and from joint ventures.

Figure 6. Organization of Robert Bosch North America in 1998, legal structure

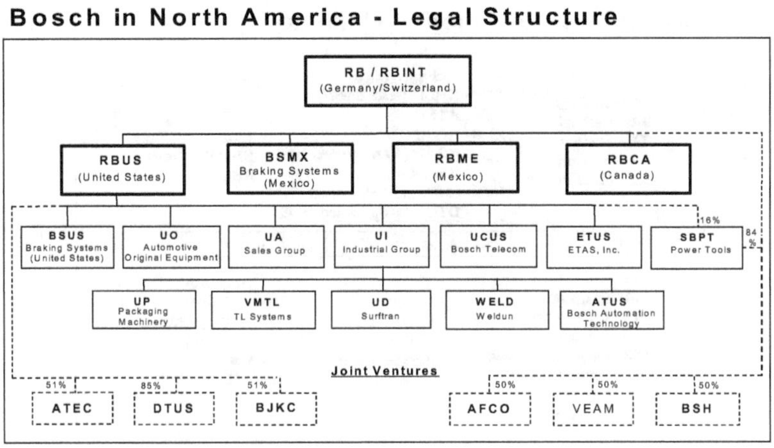

RBUS's principal line of business is manufacturing and selling various highly engineered components and systems to North American automotive and heavy-duty original equipment manufacturers (OEMs) and related aftermarkets. Major product lines include conventional braking systems, anti-lock braking systems (ABS), airbag electronics, fuel injectors, oxygen sensors, heavy-duty diesel fuel management parts, and air movement systems for automotive engine cooling and ventilation. Aftermarket products include electrical system parts such as alternators, starters and spark plugs, wiper blades, headlamps, and car audio/stereo products. The company manufactures many of these components and systems at factories in the U.S. and Mexico. The company also manufactures and/or distributes packaging machinery equipment, deburring, and cleaning equipment and services, hydraulic and pneumatic products, and communication hardware and software for the public and private network products.

Figure 6 shows the organization chart of RBUS. The top row shows that Robert Bosch GmbH (RB) (see http://www.bosch.de) and Robert Bosch Internationale Beteiligungen AG are the owners of RBUS, with headquarters in Stuttgart, Germany and Zurich, Switzerland. The second row of the organization chart shows that four entities make up Robert Bosch North America: Robert Bosch Corporation, Braking Systems Mexico, Robert Bosch Mexico, and Robert Bosch Canada.

Figure 7. Milestones for Robert Bosch in North America

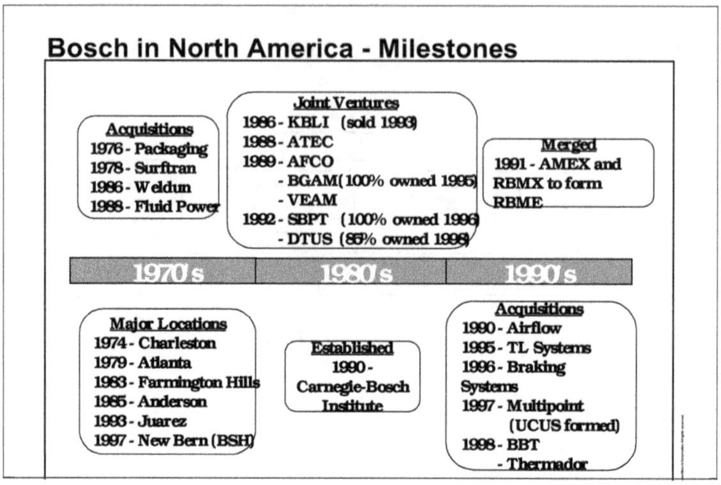

Don Chauncey described the changes that came about in the North American organization structure as follows (the milestones are shown in Figure 7):

When I joined RBUS in late 1993, there was a business called UO that sold to the North American heavy-duty and automotive OEMs. There was also an aftermarket business, UA, and there was a group of capital goods companies called UI. The latter included ATUS, an automation technology company, UP, a packaging equipment company, and others. UO was the largest business, producing about 70 % of the sales. Later we bought UCUS, which became a part of the Bosch Telecom business. ETUS, a small software development business for engine management systems was formed in Ann Arbor in April 1997. We also established several joint ventures such as Diesel Technology Company (DTUS). In April 1996 we bought the braking division of AlliedSignal, Inc., which doubled the size of the automotive business in North America and had locations in many other countries. We were responsible for taking on the acquisition in North America. The acquired business was aligned with the global K1 group that manages the braking equipment for RB. Robert Oswald, a member of the board (F4C), is responsible for K1 worldwide, as well as the North American business.

There are a number of other companies in the RB holdings in North America. These include BSH Home Appliances L.P. (BSH), a joint venture with Siemens AG; Associated Fuel Pump Systems Corporation (AFCO), a joint venture with Nippondenso, Inc.; and Automotive Electronic Control Systems, Inc. (ATEC), a joint venture with UNISIA JECS, both of which manufacture automotive equipment. DTUS, a joint venture with Detroit Diesel (http://www.dieseltech.com), is a large heavy-duty diesel business. BJKC-Americas Corporation (BJKC) is another Japanese joint venture with Jidosha Kiki Corporation, acquired with the braking business. S-B Power Tool Corporation (SBPT) and Vermont American Corporation (VEAM) are joint ventures with Emerson Electric Company in the power tools sector. This gives you an idea of the complexity of the joint venture operations. As these joint ventures and acquisitions are made, RBUS is growing rapidly and the corporate office, of which CIO is a department, is constantly adding additional services in support of the businesses. It is a very fast-moving and fast-changing organization.

When I started, there were four executives in charge of the business: Bob Oswald, Fred Hohage, Hans Weckerle, and Gary Saunders. Mr. Oswald was responsible for the automotive original equipment unit and was president of RBUS. Mr. Weckerle, who now runs the K1 business in the U.S., was in charge of the UI group as well as OE manufacturing. Mr. Hohage was president of the automotive aftermarket business. My boss, Gary Saunders, was the chief financial officer. They worked together as a management team in establishing corporate direction for the U.S. businesses. Today the NAOC (North American Operating Committee) is the top-level management committee and is made up of Robert Oswald, Gary Saunders, the President of K1, the President of SBPT, the President of UA, the President of UO, and an executive from Robert Bosch Mexico (RBME). They meet on a monthly basis to review major issues having a regional impact.

Growth of RB in North America and RBUS

Figure 8 shows how the sales of RB in North America have grown from $1.8 billion in 1990 to $5.3 billion in 1998.

The share of automotive equipment has grown from 60% in 1990 to 72% in 1998. Figure 9 shows the net sales growth of the businesses forming RB North American operations.

Figure 10 displays how the 23,419 employees of RB are employed in the different entities and business sectors, as of November 1998.

Figure 8. Net sales by business sector

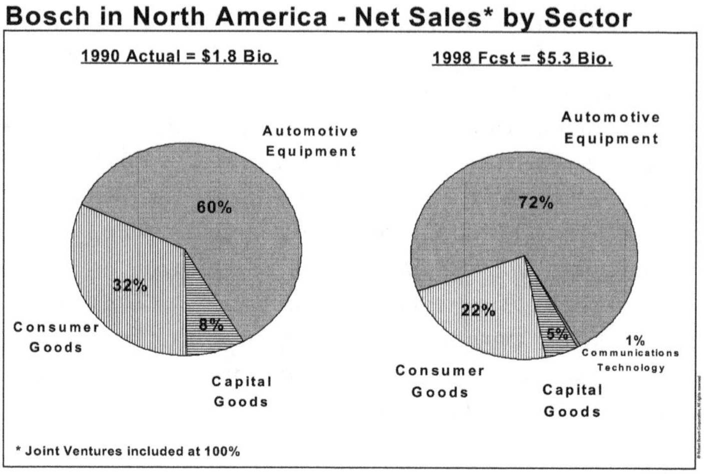

Figure 11 is an extract from the RBUS 1998 Annual Report. This figure provides key measurements of the legal entity, RBUS.

IT Organization

The information technology organization for RB in North America is shown in Figure 12. The NAOC is the top decision-making body. A Corporate Information Steering Committee (CISC) deals with strategic direction for information technology.

The vice president and CIO of RBUS is Don Chauncey. The vice president for K1/ISY2 is Mike Bieganski, who reports to Hans Weckerle. The director of CIT of UO, is Krish Kumar, who reports to the vice president of finance for the UO division. As shown, RBUS supports the rest of the organization by providing its financial, human resource, and payroll functions. The information technology organization for RBUS is shown in Figure 13. Kamran Ashraf is the director of Corporate Information Systems (CIS) and Joe DiNatale is the director of Data Services, including the data center.

Figure 9. Net sales growth by business sector

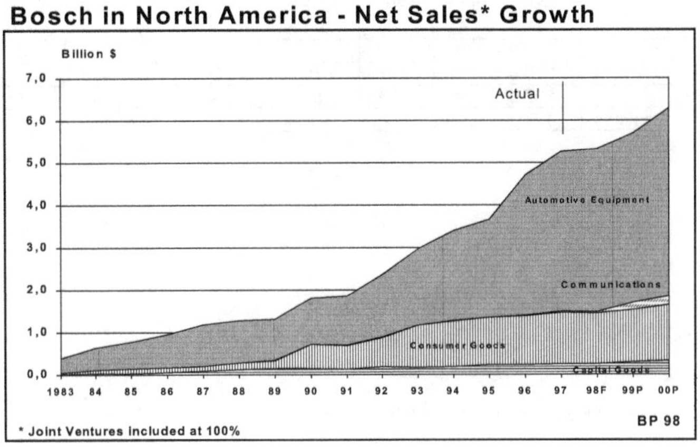

Figure 10. Employees by entity and business sector

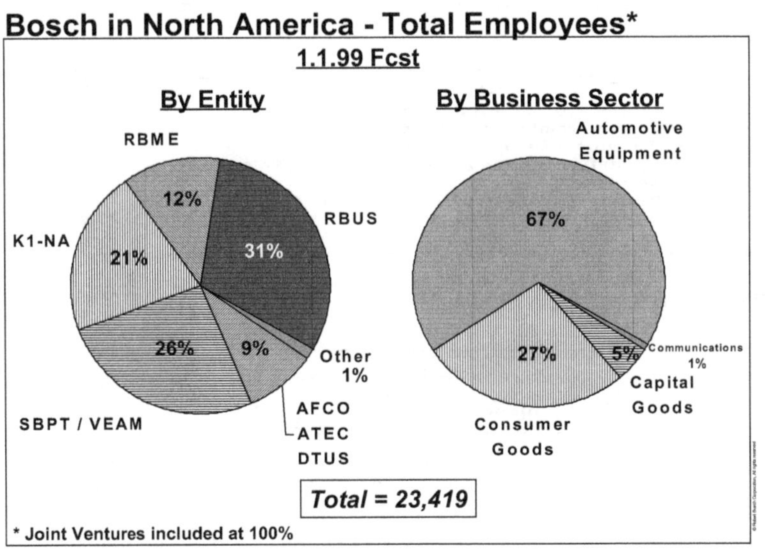

Figure 11. Selected 10-year financial data (Annual Report 1998, Robert Bosch Corporation)

in Mill.$	1989	1990	1991	1992	1993	1994	1995	1996	1997	1998
Net Sales	1,114.4	1,173.2	1,206.8	1,391.5	1,493.7	1,773.8	2,016.1	3,010.7	3,419.4	3,611.2
Income from operations	27.8	9.2	-17.8	10.7	43.2	89.2	99.0	87.7	69.6	82.3
Total assets	818.0	899.8	848.0	810.2	928.6	1,114.8	1,173.1	2,391.9	2,434.7	2,701.1
Capital expenditure	107.9	89.9	61.4	53.6	76.3	89.3	118.1	171.0	206.2	280.6
Average number of associates	4,901	5,198	5,244	5,134	4,694	4,965	5,970	9,596	11,041	12,394

Figure 12. North American information systems: IS management structure

Information Systems Management Structure

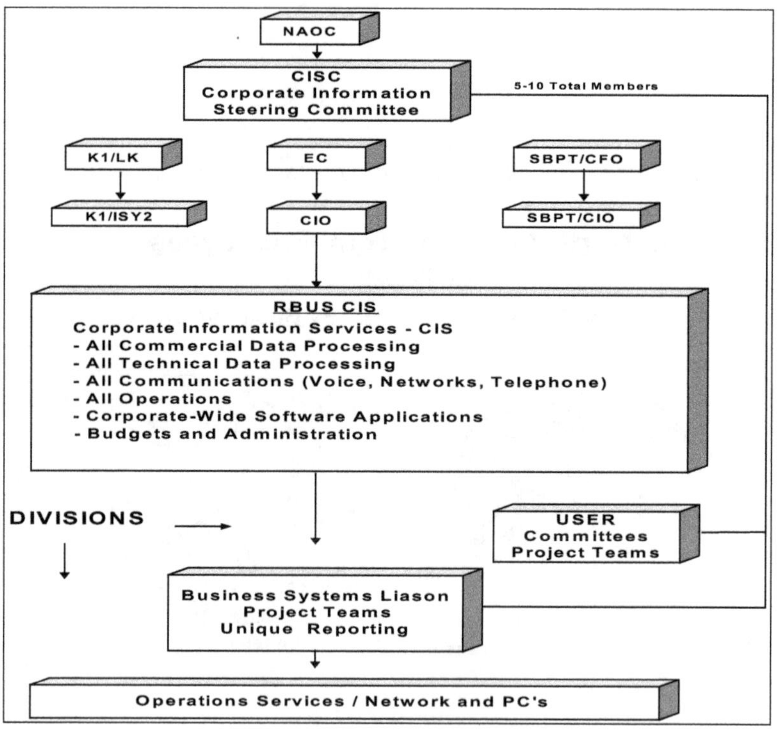

RBUS/CIO 11/98

Figure 13. IT organization for RBUS (1998)

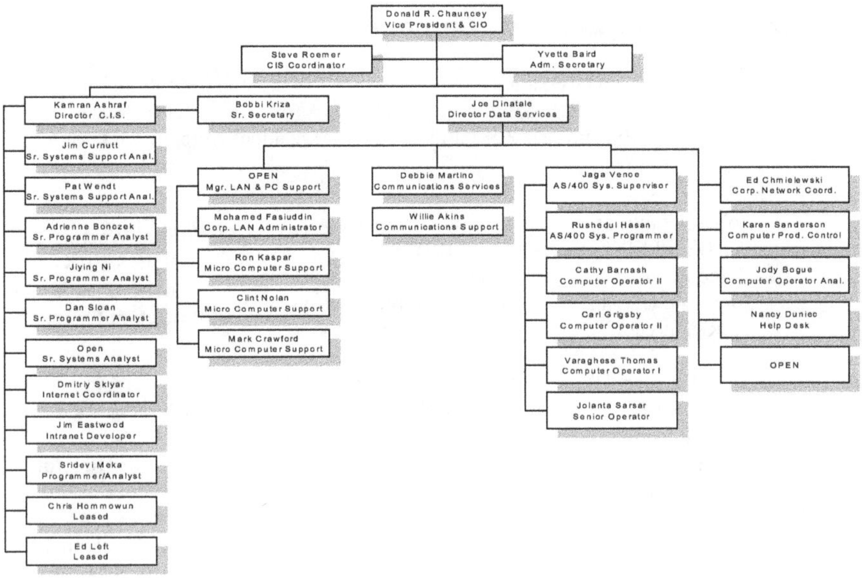

Business Structure and IT Structure of RBUS: 1996-1999

Business Structure of RBUS: Products, Processes, People

RBUS is headquartered in Broadview, Illinois. Figure 14 shows the organization chart of RBUS-Corporate. The corporate office provides financial, legal, tax and payroll, information systems, treasury, human resource, and international trade functions for the different entities in North America. Gary Saunders is the executive vice president and chief financial officer for RBUS. There are 160 associates in the corporate entity. Kamran explains some of the challenges in working for a global company:

Working in a global company, we learned that there really are differences in culture and that people pursue things differently. I think one of the things that the U.S. subsidiaries struggle with, is trying to understand the German perspective on things and why they try to do certain things vs. how they are typically done in the U.S. I think, from an IT perspective, we view the U.S.

Figure 14. Organization chart of RBUS

Robert Bosch Corporation - Corporate

	Gary Saunders RBUS/EC Executive VP & CFO	John Moulton UO President Automotive Group
Robert Cummins RBUS/PER Vice President Corporate Human Resources	Donald Chauncey RBUS/CIO Vice President & CIO Corporate Information Systems	Fred Wactor RBUS/EHS Environmental Protection Officer
Luke Baer RBUS/LEG Vice President General Counsel & Secretary	Detlef Konter RBUS/CCT Vice President & Controller Corporate Controlling	Wolfgang Knapp RBUS/FCT Facilities Director
Karen Gallivan RBUF/P President		Becky MacDonald RFBUS/COM Director Corp Communucations
Thomas Blankenship		

as a major center of innovation for systems and technology, especially in software/hardware arena. A large number of information technology innovations are coming from the U.S.

We need to improve communication with Germany. This is definitely one of our challenges. How does the organization make decisions for everybody and still make it work?

IT Structure of RBUS: Different Systems Adopted by Different Divisions

The IT applications implemented in the different divisions of North American business are shown as Figure 15. Don explains the evolution of information systems at different locations as follows:

Figure 15. IT applications in use

- **IT Applications (examples)**
  - **SBPT and K1: SAP R/3 (except F/A: Computron)**
  - **UA: J.D.Edwards**
  - **UO-Plants: PRMS (UO-Version)**
  - **RBME: PRMS (RBME-Version)**
  - **DTUS: PRMS (HR/Payroll Peoplesoft)**
  - **Financials mainly centralized: GL, F/A, A/P (Computron), A/R (MSA)**
  - **Human Resources centralized: HR/Payroll (MSA)**

When I was hired, RBUS had implemented PRMS (http://www.acaciatech.com/products/overview.html; website does not exist currently) in the smaller divisions, and I helped finish the implementation in the UO organization. Today we are running our factories on PRMS. There are some modules of PRMS that are very innovative, but it is basically an MRP II system.

The major plants of UO are in Charleston and Anderson, South Carolina. I hired Krish Kumar, who is now the director of Corporate Information Technology for UO. I sent Krish to the Charleston plant because he was a PRMS expert. We implemented it successfully. The UA division wanted software that would fully meet their requirements and selected the J.D. Edwards system that fit the distribution needs of their division.

Also, RBME was moving to PRMS from a mainframe, as had the UO organization in the U.S. So PRMS is used extensively in our company and we have significant expertise in its implementation.

PRMS was the Bosch standard here as well as in other parts of the world, and the QI standards allow Bosch businesses to remain on PRMS. However, when new ERP systems become necessary to replace PRMS, the new standard will be SAP.

The PRMS system, as shown in Figure 15, is used in the plants by UO, RBME, and DTUS. This figure also shows that J.D. Edwards is used in UA. Both SBPT and K1 have implemented SAP R/3 systems. The financial systems for RBUS were implemented on Computron software. Don discusses the implementation of Computron in his organization:

The Computron Software (http://www.computronsoftware.com) was selected prior to my arrival to replace the existing mainframe financial applications. It has been used for our operations on the corporate level for a number of years. It's a very innovative system. It includes workflow functionality, and allows scanning and indexing of documents such as invoices. On the AS/400, Computron is running G/L, F/A, A/P (General Ledger, Fixed Assets, and Accounts Payable). We completed the conversion to Computron in August 1998. In the process, we eliminated some old systems and the significant costs associated with them. We also eliminated a DOS/VSE mainframe and its cost.

The conversion process began about six months after Don joined the company in 1993, and was complete by August of 1998. Many people were involved and were faced with significant change. Systems people ran the project. Don concluded the description of the current systems by stating:

We were driving this, and accomplished it credibly because we understood the business process and were able to put all acquired companies on to our systems. We are now almost entirely on AS/400 systems, including PRMS, Computron, and J.D. Edwards. The only mainframe systems remaining are the Human Resource/Payroll and the Accounts Receivable Systems running on an outsourced mainframe.

As shown in Figure 15, RBUS outsourced the operation of payroll application to May & Speh while maintaining the software support internally. The same applies for A/R (Accounts Receivable), where an MVS mainframe was used to process the payroll application. All payrolls for RBUS were done using this system except DTUS, which ran its payroll on the PeopleSoft system and SBPT, which used a full outsourcing provider.

Business Structure of UO: Products, Processes, People

UO is headquartered in Farmington Hills, Michigan. Krish Kumar, the director, CIT for UO describes the products and processes in UO as follows:

UO is an RBUS division and has primary responsibility for all sales of K3 to K9 product lines in North America. UO is about $1.5 billion in sales, with four major plants (see http://www.bosch.de/k/englisch/atlas/atlas.html) in Charleston and Anderson, both in South Carolina, Juarez, Mexico, and Albion, Indiana. Next year, RBME, in Toluca, Mexico, will become part of our division producing K4 and K9 product lines. Our growth has been significant during the last few years. DTUS, part of the diesel business, will also become part of the UO division, in the near future. From the information systems perspective we have brought DTUS onto our systems already, but they will be integrated from a business perspective in the future.

IT Structure of UO

The IT division at UO has approximately 59 people and reports to Krish Kumar. Krish explains the IT structure of the UO organization:

I am in the headquarter's site of the UO division. I have a small team based here but we work very closely with Don's group in Broadview, with the rest of the plants in UO, and also with Mike Bieganski because we also support the braking division people who work in Farmington Hills. With respect to products, we have two lines of business. About 60% of the business is produced in our factories and 40% is purchased "margin business." When I say margin business, these are products that we import from Bosch plants in Germany, Brazil, India, Turkey, or other countries where Bosch plants are located. The value added here is based upon fulfilling customer needs with products produced in our factories, or by directly shipping the products to our customers from plants around the world. There is significant complexity in the latter, as the products come from our internal supplier to our warehouse and are then shipped directly to our customers. We are responsible for the inventory, the delivery of the goods to the customer, and the engineering changes that occur, even if we do not fully control the manufacturing of the product. There is additional complexity; if because of capacity constraints we are not able to produce a product, fuel injectors for example, we need to import it from Bamberg in Germany for shipment to our customers. In that case, we should be able to handle the order as a single product, whether it is produced in our factories or purchased from elsewhere.

Currently, we are running the PRMS system for our manufacturing operations. We tie into Broadview for the financial systems. We also have in-house written systems that address the shipping, billing, and invoicing to the customer. We have a system that ties EDI (Electronic Data Interchange) both to our customers and our suppliers. We have about 11 systems that address our total business cycle. The demand management system determines what we have to buy based on engineering data. Therefore, we tie the systems to the engineering piece of the organization, for example, the routing and bill of materials for manufacturing the product. Then we build the product using manufacturing methodologies. To build the products we need components and source these components from our domestic and foreign suppliers. After we have built the product, it is shipped directly to our customer. This involves maintenance of shipping, invoicing, and billing systems. Then we build a financial summary for internal reporting purposes as well as for our external reporting at the end of the quarter, and give it to Robert Bosch in Germany.

Krish further explained about the EDI system[i]:

Ninety percent of our demand management is done using EDI with our customers. We have 38 OEMs, and each OEM might have 9 to 10 different plants where we deliver our products. We conduct EDI transactions with the top 135 of our suppliers. So you can see how tightly integrated the EDI system is to the manufacturing and demand management functions. We also use the EVA system (the order entry system in Germany) for demand management of international requirements.

The other critical functions of our manufacturing systems are at the Charleston and Anderson plants. These include capacity management, production scheduling, production optimization, and flexible team manufacturing. For example, one morning we might get a material release of a customer requirement. At the same time, the customer will give us a ship schedule, expecting us to manage the change effectively. We can sign onto the customer's system and see their production at any given time. So we have to combine the demand, the material release, the ship schedule, and the capacity utilization of their production. That's where the production business and margin business come into the picture. If we do not have adequate capacity, we have to purchase the product from other Bosch operations to meet the customer demand. We also use the EMS system, which creates documentation for U.S. Customs for the export of manufactured product and noninventory exports.

Challenges Facing RBUS/UO

During 1998, Don and Krish were faced with many challenges, as the multiple information systems were not able to meet the needs of the growing and changing business at RBUS and UO entities.

Challenges Facing UO

Krish explained how the different systems made it difficult to integrate the operations and meet customer demand effectively:

To manage the customer demand precisely, we have a very sound physical layer, we have a very sound network layer, but we do not have a very sound application layer. That's what we need to focus on in the future to meet our objectives.

Because of the current nonintegrated system solution, we have created pockets of implementation gaps between the functional departments, sales, engineering, manufacturing, distribution, logistics, and financials. Not only do we have complexity in the organization, but also the old reporting methods add time to the production of performance information. So management requires us to position the application layer to address business growth and increased complexity.

Further, our customers are asking us to provide both total systems and modules. They want to reduce the number of suppliers. This brings more complexity. We have to work closely with the OEMs. The whole system or module is integrated into the automobile, and we have to bring in our tier-2 and tier-3 suppliers, and even our competitors at times. We need to ensure that the design integrates with the other components of the car.

RBUS HR Requirements

Don talked about the issues requiring changes to the payroll and human resource system:

Though we had changed every other system in the company in five years or less, we had not changed the payroll/human resource system. In fact, our payroll system is an old mainframe based system. It has a human resource component, but the users indicate that the system doesn't have the human resource capabilities that are required. The vice president of corporate HR wants very urgently to implement a new human resource system. About a year ago he requested a study for a new system.

After discussions with Computron, they finally agreed that they would support the AS/400 version of their software until the end of 2001. They have other customers who had Y2K issues and they want to support their customers through Y2K so as not to further add to the workload.

Alternatives Considered by Krish and Don

Krish wanted to align the information systems in UO with business processes in order to support the requirements of his end-users. He stated:

There are many projects ongoing already here and in Germany. If you look at this complexity there is no way that I can address the business requirements with the existing systems. The only way I can accomplish this is with a clear strategy of understanding the future business processes and aligning the system configuration accordingly.

Exploiting the information systems is where I see the benefit of the entire investment for Bosch in IT. Take for example the process of receiving an item, storing it, moving it to an assembly, and understanding the transaction costs of this process. Today we use multiple systems for these steps. With the current systems, I can't tie together some of the critical pieces of information. The existing systems are not aligned with the automotive vertical market. The result is that within the IT organization we are fighting fires constantly. The software does not accommodate the growing and changing business requirements, and we have to tell our customers that the system can't deliver the information they want.

Our end-users are asking for real-time information. I cannot have the IT associates function as program generators and developers anymore; they have to become business-minded people, helping their users by actively working with them and resolving their business issues.

Don had to consider the withdrawal of vendor support for the existing financial systems on the AS/400 platform. He was also looking at PeopleSoft for the human resource system replacement, since at that time SAP had just released its Payroll/HR system in the U.S., and there were only a few installations. The human resource department, along with Kamran, did an analysis of various packages. Don explained:

At that time, PeopleSoft was the leading HR/Payroll software in the U.S. market, and several of the human resource personnel in our divisions had prior experience with this software in other companies. We therefore considered Lawson (http://www.lawson.com), SAP, and PeopleSoft. Our people were leaning towards PeopleSoft, but I had concerns whether Germany would accept this decision. At that point I met with Dr. Eggensperger at QI-Daysii in 1998. He said that in order to make a decision on this issue he would need a formal proposal.

Don had to establish a strategy to make multiple systems work together in order to cope with the growing and changing business environment.

Chapter Summary

Robert Bosch GmbH follows a traditional corporate structure of four main business units that are overseen by the Board of Management. In 1993, problems with legacy systems forced the RB management to reexamine the IT structure of the company. In order to implement new missions, a new division by the name of QI was formed. Dr. Eggensperger, Mrs. Bauer, and Mr. Mierzwa supervise and manage the three units within QI.

Don Chauncey, vice president and chief information officer (CIO) for Robert Bosch Corporation (RBUS), had talked with Computron Software, Inc. executives during 1998. They told him that they were going to support their financial systems software on IBM AS/400 computers only through December 2001. In the future, Computron wanted to provide these functions on a client-server platform. Don faced an additional demand from the RBUS Human Resource (HR) function. The department was requesting applications that were superior to the current systems, which were operated on an outsourced IBM mainframe computer. The HR and IS staff recommended to Don that PeopleSoft be used to address the payroll/human resource needs.

Krish Kumar, Director of Corporate Information Technology (CIT) for the Automotive Original Equipment Division (UO), was also evaluating the current systems in order to address customer needs. The current nonintegrated system solutions had created pockets of implementation gaps between the business entities. Not only had it created complexity in the organization; but the old reporting methods also added time to the production of performance measures. Furthermore, the trend of the Original Equipment Manufacturers (OEMs) was to have their customers deliver total systems and not just parts.

One of the major issues QI was concerned with in 1994 was the implementation of an ERP system that fulfilled the requirements of the Board of Management. To better understand the underlying process within RB, this chapter takes a closer look at the North America operation. RBUS's main focus is on the automotive industry and large OEM clients. During 1993-1998, RBUS has gone through a period of acquisitions that increased sales by $5.3 billion and led to a growth of the company. The period of growth reduced the ease of personal communication as well as technical inoperability among the different divisions with RBUS. The company had implemented multiple information systems that did not communicate with each other seamlessly. Don had to establish a strategy to make the multiple IS demands work together in order to cope with the growing and changing business.

Team Assignments and Discussion Questions

1. Krish Kumar stated: "I cannot have the IT associates function as program generators and developers anymore; they have to become business minded people helping their users by actively working with them and resolving their business issues."

   (a) Analyze and evaluate the organizational structure of Robert Bosch GmbH and Robert Bosch North America considering organizational theory.

      (i) Describe and analyze the business structure of Robert Bosch GmbH.

      (ii) Describe the structure of Robert Bosch GmbH and the structure of Robert Bosch in North America

(iii) Analyze the structure using organizational theory for global companies

(b) Analyze and evaluate the interdependence and connection between organizational and/or business environments and the information systems structure for Robert Bosch North America. Consider corporate and divisional responsibilities in the IT area.

2. Assume the role of Don Chauncey and propose solutions that address the challenges faced by the UO, RBUS financials, and RBUS HR functions.
3. Assume the role of Krish Kumar and propose solutions that address the challenges faced by UO functions.
4. Basically, there are two diametrically opposed possibilities. One is to stay with the existing systems and try to optimize them, the second is to follow the corporate guideline to go with the SAP R/3 system. Summarize the advantages, disadvantages, issues, and possible strategies for:

    (a) staying with the existing systems and

    (b) implementing SAP R/3 throughout RBUS.

References

Alliedsignal. (1996). Retrieved September 10, 1999, from http://www.alliedsignal.com; http://www.alliedsignal.com:80/corporate/media_kit/press_releases/1996/feb29_96.html

Bosch GmbH. (1998). Retrieved September 10, 1999, from http://www.bosch.de/k/englisch/info/info.html

Endnotes

i UO uses an EDI system from future3-software, see http://www.future3.com/index_d.html); Web site nonexistent currently

ii Worldwide conference held every year for the IT people from the Bosch group

Chapter VII

"Unrest Produces Creativity": Standardization of Business Processes in Robert Bosch Corporation with the Use of SAP R/3

Learning Outcomes:

- Understand in detail how an ERP system can be implemented in differing ways in a company, with resulting benefits and problems
- To show with a detailed industry example how if an ERP software is not implemented correctly, it may lead to multiple ERP systems existing within an organization that cannot communicate with each other
- Understand decisions that were made at Robert Bosch GmbH and RBUS regarding SAP R/3 implementation between 1992-1999

Case Overview

During August 1999, Don Chauncey, Vice President of IS for Robert Bosch Corporation (RBUS), finished making his presentation on the direction of an SAP implementation to the RBUS North American Operating Committee (NAOC). The plan was to introduce SAP R/3 Financials, Human Resources and Payroll, and Logistics modules for RBUS and UO (North American Automotive Original Equipment). He had complied with QI's (Querschnittsbereich Infomations – verarbeitung – Division for Information Technology of Robert Bosch GmbH) request to define a project for the implementation of SAP in the North American Bosch businesses.

Before proceeding further, the NAOC requested additional information regarding the project costs, specifically how much was due to: (1) "must do" items, (2) a corporate standard, and (3) the need to improve processes. The NAOC also wanted to know what the current running costs were vs. the forecast costs once SAP had been implemented. The critical concern of the NAOC was the overall cost of the implementation of the SAP system.

This was a major challenge for Don as he started work on revising the proposal. In Germany, QI had established SAP as the standard software to be used within Robert Bosch. Mike Bieganski, the vice president for Information Systems and Services at the K1 Division, presented possible alternatives that could be adapted to address the concerns of NAOC. Krish Kumar, the director of Corporate Information Technology for UO, was certain that implementation of SAP R/3 in his division was imperative. Kamran Ashraf, the director of Corporate Information Systems, and Don started planning their strategy to address several major issues:

a. How to structure the SAP system to keep it flexible and ensure consistency,
b. How to design the implementation process,
c. How to justify the cost of the SAP implementation from a business standpoint.

QI's Request for Common Systems Globally

Establishment of QI

Reviewing the problems with legacy systems in Robert Bosch, the Board of Management defined two mission statements for the IT management team during 1994. First, the IT division had to operate like an external IT provider, and meet or exceed the benchmarks of these providers. Second, the division had to be a leader in setting policies, procedures, and standards for IT throughout the organization. The board gave the division a chance to change its own environment and establish its own future, rather than have an external consultant come in and recommend changes.

Until now, the different locations within Robert Bosch GmbH had used their own systems, which they had optimized to suit their own requirements. They had their own little paradises. They did not see any need to change their system to a new Enterprise Resource Planning System (ERP). However, there was external pressure from the market for the units to work together, and too much time would be needed to make this possible with the old systems. In addition, if the company wanted to purchase new companies, a much more flexible system was essential. An application architecture was preferred that started from the top of the organization (board) and went to the bottom (shop floor).

The IT management team created the QI division in order to address these requirements. Robert Bosch began using SAP R/2 for accounting in the early 1980s, and the board decided that implementation of an SAP system across the organization would address the requirements adequately. In different plants and

Figure 1. Time line showing fully implemented SAP R/3 projects at Robert Bosch

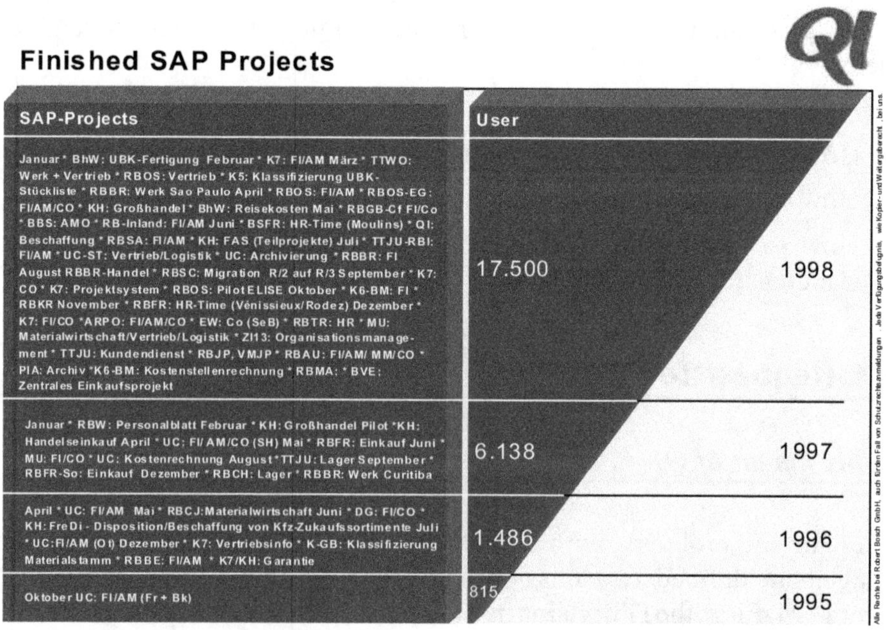

locations across many countries, joint teams initiated SAP R/3 implementation. During 1995, FI/AM systems (Financial and Asset Management) were implemented, serving 815 end users. As Figure 1 shows, the number of end users increased to 1,486 by 1996, when nine systems in different financial units were implemented. In all, 17,500 end users were using more than 50 SAP R/3 systems across Europe by the end of 1998.

As a result, Robert Bosch was in the position of having many locations in Europe that were using SAP R/3, but the systems had not been implemented uniformly. The board told the IT management team that the current implementation was not acceptable since it did not meet their business requirements. They wanted QI to provide the ability to close the books for all Bosch operations and report the results within 5 days.

QI-Mission

Since 1995, the European IT activities of the Bosch group had been managed by the cross-functional information processing department (QI). QI supported the automation of processes and the engineering of business processes, as well as the provision of information in commercial and technical areas. The mission of QI was to increase the productivity and improve the competitiveness of the company.

The use of IT in the Bosch group was oriented to the economic and safe operation of IT systems and the introduction of new IT technologies to improve the company's competitive positions in internationalization and optimal use of distributed resources. Considering this, the QI goals were:

- customer orientation,
- standardization,
- efficiency,
- planning for the future.

QI provided the IT infrastructure, the computing centers, and networks (both local-area and wide-area), as well as terminals (PCs, workstations, and other IT devices) to the end users. Another task of QI was to configure the standard software (e.g., SAP R/3) to RB's business needs. It was also responsible for the development of software packages that were not commercially available.

QI helped its customers cope with the constant change and technological progress by the innovation of specific solutions. This required them to adapt hardware, software, and procedures that were versatile and met "industry standards." Guidelines used by QI in order to make this work were:

- standard software from outside vendors preferred over internally developed software,
- use of existing hardware, software, and process solutions in multiple divisions,
- extend and cross-link the IT infrastructure across the organization,
- provide enterprise-wide support for all business functions locally,
- use of bulk-purchase potentials,
- initiate innovative projects to meet future requirements.

A substantial part of QI dealt with the development, implementation, and maintenance of application systems. New concepts and their development were as important as consultation on and maintenance of current systems. Finally, they were responsible for creating the system architecture of the future. QI was responsible for a wide range of different kinds of applications, including:

- *Business applications*, such as purchasing, payroll, or accounting, had to take into account the challenges of the next century. For example, the introduction of the Euro, a common currency in Europe, had a significant impact on commercial applications.
- *Technical application systems*, such as Computer-Aided Design (CAD) and Product Data Management (PDM) in the mechanical or electronic engineering areas, had to be integrated with other IT systems.
- *Sales applications* that focused primarily on the wide range of customers of the Bosch merchandising business had to be able to provide information on a particular customer for the whole company. This area was characterized by a large number of products and inventory management needs.
- *Plant applications* that needed to support the entire supply chain. These had to provide both German and foreign plants with complete solutions using a combination of SAP R/3 and supplementary subsystems.

In addition, QI was responsible for supporting the applications and IT units for all the Bosch companies and joint ventures around the world. This responsibility included the management and coordination of projects, as well as the realization and marketing of standards in both the hardware and the software areas.

SAP R/3-Strategy

A press release from the SAP AG, Walldorf, on July 16, 1999, gives a good description of the way RB began to implement SAP R/3:

In a period of just over four years, the number of R/3 users in the Bosch Group has risen from almost none to 20,000. The 20,000th R/3 user to go live works at Bosch's automotive equipment/bodywork electrics division in Belgium. In the medium term, over 50,000 users at the world's largest independent automotive supplier could be using SAP's market-leading integrated solution to manage the company's business processes.

Bosch began using SAP R/3 at the start of 1995, in its telecommunications division (Bosch Telecom) in Frankfurt. Today, Bosch employees all over the world work with R/3 every day in all business units — including automotive equipment, power tools, household appliances, thermotechnology, automation technology, and packaging machines, thus emphasizing Bosch's decision to make R/3 its software of choice for mapping and supporting its business processes.

So far, Bosch has been using SAP R/3 in the areas of Accounting, Logistics and Human Resources, deploying practically all of the classic R/3 modules. Around the world, Windows NT is used as a server operating system in several "server farms" (about 150 servers in total). Bosch currently has over 40 R/3 systems. IT projects instituted more recently are based on products in SAP's New Dimension initiative, namely the SAP Advanced Planner and Optimizer (SAP APO), SAP Business Information Warehouse (SAP BW), and SAP Business-to-Business Procurement (SAP BBP) in conjunction with R/3.

Within its cross-functional information processing department (QI) Bosch has a dedicated internal team of consultants who deal specifically with R/3 implementation and use. The pool's members combine specialist knowledge of R/3 with general business know-how. They cover not only all relevant SAP topic areas, but also ensure that standardized master data, basic structures, and templates are used. Besides carrying out project work and safeguarding Bosch's process knowledge, the team employees ensure a continuous know-how transfer between projects, locations, and business areas and between Bosch and SAP. The team was involved in the R/3 project in Belgium.

Figure 1 shows the implementation of SAP/R3 across Europe and how it has grown from supporting 815 users in 1995 to 17,500 users in 1998. Mrs. Bauer, who was responsible for the software applications unit in QI, explained the past approach in implementing R/3:

We initially designated the new standard as SAP R/3. Everywhere, where there was a need for a new application system, we started a new SAP-project. As a result, we ended up with many locations in Europe using SAP R/3, but none of the systems are the same. This was not what we wanted, so we have changed our approach from bottom-up to top-down. We are now developing a common product for use in every plant location, so the processes will be the same. We hope that this approach will be successful.

We are not doing this in a closed forum. We have included people on the project team that will standardize the configuration. We have five pilot plants and about 170 people involved in defining the processes. To harmonize processes, there will have to be changes. Not everyone's processes will fit. A benchmark will be set in the end. The experts in these areas will determine the processes, and communicate with the various plants. By the time it is finished, everyone will know what is coming and what will be in the product.

To achieve improvements we are trying to make the end users think about their future business needs, and to really change by establishing measurements. For example, we have set goals for the ERP project that every process is 30% faster, or has 30% fewer steps. We want to establish a very formal process with SAP to establish KPI (key process indicators). That means that we need strategic goals from the board of management, like achieving good customer relations. Once we get those, we have to break them down into operational goals. We then must establish measurements to show process improvements. Before a new process is accepted, it should be simulated to see if the measurement is improved. New processes will be accepted only if they improve performance.

We just started this process on July 1, 1999. We are using a procedure model that is similar to the SAP R/3 model, but we have changed it a little bit. The first step is the study. It will end with a prototype of the product in February 2000. We will have defined all of the processes and documented them. It will take another three months until the final product is developed. It will then be implemented in five pilot plants. In the second quarter of 2001, we want to start the rollout to our other locations.

QI Standardization

QI created a project to define and to use a common system throughout the RB plants in the automotive equipment business sector (UBK – Unternehmensbereich Kraftfahrzeugausrüstung) and the EW division (Elektrowerkzeuge – power tools). Mrs. Bauer stated in the QI-Newsletter of July 1999 that standardization is one of the most important goals within QI:

Standardization is the focus of our attention these days with the RB/3W project, one of the largest projects in the history of Robert Bosch GmbH. RB/3W is the standardized IT product for UBK plant logistics, developed for and by Bosch on the basis of SAP R/3.

"What's this now? Just another thing that makes life more complicated!" I could hear one or the other person moan; "Our plant is already well optimized!" True — for the individual site and for today's requirements. But what about the company-wide situation, and are today's processes the right ones for our future success? Finally, I would like to define the term "standardization," as I understand it. For me, standardization does not mean leveling down differences. No, standardization means getting together with the experts to determine and introduce the best process for a certain procedure (e.g., the kanban system) or the best system for a certain product (e.g., document management).

1999: Change as the Opportunity for Development

The QI Newsletter of July 1999 reported more details of the RB/3W project:

If unrest produces creativity, as QI's Jürgen Schlaefer defined his personal motto during the "QI Days" in Karlsruhe, Germany, there is a good chance that the RB/3W project will be marked by creativity. Preparations for the project, which, according to an external partner, is the most comprehensive in Bosch project so far, have led to unrest in the positive sense of the term on all company levels.

It is not only the motto "Creativity through unrest" that is unusual. The aim of the project, the replacement of the current mainframe systems, represents a challenge. Among them are PUMA (Production and Material Planning System, formerly EDP-PPS), which is used in the Feuerbach,

Hamburg, and Reutlingen plants; R/,2 which is used in the Bamberg plant; and self-developed systems on the BS2000 platform used in the Bühl plant. What do these five plants have in common? These plants are the pilot plants of the RB/3W project managed by Jürgen Schlaefer (QI/ASW). The project team consists of employees from the plants and QI and has been assigned the ambitious task of replacing today's applications. What will be replaced is a landscape of systems that meet the current requirements but are costly and use more resources. In the age of globalization we have to fulfill requirements such as:

- *international cooperation (production and development alliance, Global Sourcing, etc.),*
- *electronic links for customers and suppliers (e-business),*
- *flexibility in assigning business units, and*
- *modern, easy reporting (example: "What was the total turnover Bosch produced selling product A to customer B?")*

Today's IT systems, which are meeting requirements in the plants well, are now outdated. The main reason for renewal is, however, the need for standardizing our processes throughout the company. Our primary goal is to remain the strategic partner for our customers. And that implies shaping our internal and external processes so that we can act fast and purposefully, at the lowest possible cost.

The RB/3W project team will develop an application system that will be the standard product for all K-plants (i.e., in Europe, for the time being). RB/3W stands for the introduction of a standard product on the basis of SAP R/3 to support UBK plant logistics at Bosch. Project organizations play a vital part in realizing the RB/3 plant systems. Various monitoring and consultative bodies, such as the User Control Team and the Steering Committee, ensure product development that facilitates standardized processes and guarantees the scope of functions and their viability. It is thus ensured that existing procedural knowledge from today's applications and the targets for the future are assessed in the same manner. The Steering Committee ensures that the goals of the project can be achieved. It consists of members of the Bosch management (Messrs. Hoffmann and Dais), the commercial executives of the K-division's pilot plants (K3, K4, K5, K8), various central departments (ZE, ZI, ZQ), and QI/LA. To be able to make the necessary resources available to the project and thus achieve the ambitious project aims, the management decided to freeze all existing systems, as well as any R/3 plant projects other than RB/3W. Indispensable further developments of the current applications must be approved by the Steering Committee. The project goals are at predefined costs and within the approved time schedule:

- *define and develop the RB/3W product,*
- *introduce the product in the pilot plants, and*
- *prepare and implement the general introduction in all other plants and replacing existing systems*

Ambitious milestones have been defined: By the end of 1999, QI/ASW, together with the pilot plants, will have prepared the specifications, that is, the standardized and documented procedural scope of the RB/3W product, and will have proved the efficiency. After that, in February 2000, the management will decide the release for realization. Introduction in the pilot plants is scheduled for as early as July 2000. After the successful conclusion, planned for March 2001, the system will be released for comprehensive introduction into 63 plants in Germany and abroad. The development of the RB/3W product is a Bosch-wide corporate project. The management, the central departments, plant managers, the departments, and all QI units support it. The major difference, as compared to other R/3 projects at Bosch, is that first a standardized product (RB/3W) will be developed, which will then be introduced in all plants concerned. The economic efficiency of RB/3W will be measured against the overall benefit for Bosch. If necessary, work sequences realized in the current systems will have to be adapted in some areas in the course of standardization. RB/3W is a project of the user or functional departments, not of QI, because it is primarily concerned with simplifying and standardizing business and manufacturing processes. We would like to encourage the departments and developers to be sensitive and creative! So, let us turn these changes into opportunities for development and for making provisions for the future! If we do not take our time to shape our future today, we will have plenty of time in the future to think about our past.

Development of Proposal by Don: 1998-1999

Don Chauncey, as the vice president and chief information officer for RBUS, had to develop a proposal for implementing SAP R/3 for the RBUS operations. He had to consider cessation of support for the existing financial systems and potential use of PeopleSoft for the human resource functions. He also wanted

Figure 2. SAP R/3 implementation and costs

SAP R/3 implementation

costs: Create transparency for internal customer

- It is efficiency that is essential for QI not the total costs nor the number of employees.

to take into account the request from Krish to use SAP R/3 in UO. He was well aware that QI wanted RBUS to implement SAP R/3, as shown in the slide in Figure 2 that states with respect to costs, "It is efficiency that is essential for QI, not the total costs nor the number of employees." The directive from QI was clearly telling him that he had to implement SAP R/3 in RBUS and UO as part of the effort to standardize processes all across Robert Bosch.

The details of the process he used to develop the proposal are discussed in this section, and an overview of the process is as follows:

- During 1998, RBUS created a proposal for implementing HR solution with PeopleSoft. It was not accepted by QI, who asked RBUS to implement this function using SAP R/3. QI agreed with Don that FI should also be implemented.
- During March 1999, Don started work on a proposal to include FI/CO and HR.
- During August 1999, he worked with Krish Kumar to include the logistics function of UO.
- During October 1999, Don and Kamran started revising the proposal so that it could be presented to the NAOC, thereby making it possible for the SAP R/3 implementation to proceed at RBUS.

1998: Suggestion to Implement HR Solution with PeopleSoft

To replace the legacy mainframe-based payroll/HR system, the HR department first proposed to implement the PeopleSoft Human Resource System as their preferred system. At that point, Don met with Dr. Eggensperger at QI-Days in 1998, which is an annual event where key representatives of the IT personnel employed by the Bosch Group meet. He discussed with Dr. Eggensperger the proposed direction of going forward with PeopleSoft and it was agreed that he would put together a formal proposal. Don explained his approach:

I wanted to move forward with a proposal as soon as possible. QI had already indicated that we could not go forward with the client server version of the Computron Software, Inc. (Computron) financial systems, and I felt certain that SAP would be the choice for HR/Payroll as well.

Prior to 1998, there were changes in two divisions: Skil Bosch Power Tool (SBPT) and K1. SBPT was running on a legacy system that had significant Y2K issues. At the same time, we bought the braking division of AlliedSignal, Inc. They had software running on a mainframe for which they did not have the source code. As a result, they were not able to add or change any of the source code to make it compliant for Y2K processing. They had to replace the existing systems. The braking division had planned to implement the QAD solution (see http://www.qad.com), which was a good automotive solution in the U.S. But after the acquisition, QI reviewed this decision and indicated that they must move forward with SAP R/3. So both SBPT and K1 got project approvals to implement SAP.

Though the RBUS systems are Y2K compliant, there was significant demand to move forward with HR/Payroll and a need to replace the Computron financial systems due to Computron's projected withdrawal of support for AS/400 platform. Therefore, I began a project to get the approval for the implementation of SAP R/3 Financials and HR/Payroll.

March 1999: Include FI/CO and HR

In February 1999, at a meeting between RBUS/CIO Don Chauncey and Mrs. Bauer, QI/LA, it was decided that an SAP R/3 project including FI/CO and HR would be implemented. One reason for starting with FI was the fact that HR generated postings to the General Ledger (part of the FI function). If this was not integrated, it would lead to significant problems in creating interfaces for the FI function. By starting with the FI project, standardization could be implemented on the Bosch chart of accounts for companies in Robert Bosch North America. Two FI experts from QI, Mrs. Reinhardt, who had implemented the systems in the UK, and Mr. Bauer, who had done a similar implementation in France, supported RBUS in preparing a project request and estimated project costs during March 1999. At the same time, QI members from the HR area were in the U.S. for one week. Both FI and HR suggestions were reviewed with experts from SAP. Also, Mr. Schlaefer and Mr. Kleiser from QI were in the U.S. regarding supporting the logistic processes with SAP R/3. Don selected several consulting firms that could work with RBUS as external partners on the proposed implementation:

We participated in SAP's SAPPHIRE Conference a year ago and brought in four consulting firms: Andersen Consulting (http://www.andersonconsultinggroup.com), Ernst and Young (http://www.ey.com), Whittmann-Hart (www.whittman-hart.com), and CCAi (http://www.ccai.com). We selected two large accounting firms and two smaller firms. All of them had won the SAP award of excellence in the previous year. We brought them in and we went through all the processes. In the final analysis, we narrowed the field to Andersen and CCAi.

In March 1999, a proposal for the North American SAP Project which included HR and FI was presented to the RBUS and division management. The following divisional assumptions were included:

- K1 would not initially be integrated into the RBUS Financial SAP System
- A shadow (G/L & F/A) company would be implemented for consolidations
- SBPT would not be converted from the Computron Fixed Assets system to the RBUS SAP F/A system, but would implement the SAP F/A on its own system
- Initially, SBPT would not participate in the RBUS implementation of SAP HR/payroll

The allocation of funds to the SAP/R3 project conflicted with demands from internal projects and could have a serious impact on division profitability.

August 1999: Add Logistics Function of UO to the Proposal

Krish, the Director of CIT at UO was enthusiastic about implementing SAP R/3 in his division. He did not believe that SAP R/3 would be the magic bullet providing a complete solution, but it was an opportunity to address the missing application layer in the IT-architecture of UO and RBUS:

To manage the customer demand accurately, we have a very sound physical layer, we have a very sound network layer, but we do not have a very sound application layer. That's what we need to meet our objectives.

The major impact that I see in the organization that would help us in the application layer is to address the critical business issues from sourcing, to engineering, to manufacturing, to distribution, to financials, and to logistics. And one component that is very critical is quality. We have several island solutions that are being developed to address the quality needs of the customer base and quality needs of the supplier base. All of these are the components of the application layer that could add a significant value through the implementation of SAP. Then we can focus on the presentation layer. This could be addressed by business warehousing, where we can provide accurate and timely information to make decisions by the entire management and we can share the data with our customers and our suppliers. The other factors in the automotive industry, that is, quality, cost, and delivery, are pretty much given; you can't deviate from industry standards. So what is left is only one component, and that's time. And that is the only piece where the IT organization can increase the value added by providing timely information and accurate information to our internal and external customers.

The increasing margin business within UO made it necessary to harmonize the business processes between the plants in the Bosch group. Krish explained the advantages of having a standardized application architecture with an example:

Let us take the example of the fuel injector produced in Charleston, which is the same as one produced by our Bamberg, Germany plant. If the product planner in Charleston, South Carolina, can sign on and see the same applications in the two plants, we can increase his effectiveness tremendously. He might notice the fluctuations in demand. He might notice that the Charleston plant lacked capacity to address the demand here, and schedule the line two days later. Perhaps this product is being produced this afternoon in Bamberg, which can therefore become a second-tier supplier. What has he done in this process? He has reduced the inventory, carrying cost across the supply chain. In addition, he has scheduled and optimized not only our delivery to the customer but also the delivery of the product from Bamberg to Charleston. A prerequisite to make this happen is the alignment of the processes between the different Bosch plants.

From a service-provider point of view, Krish as an IT manager saw a significant potential for improving the company's business processes using R/3.

August 1999: Presentation by Don to NAOC

Don incorporated the logistics functions in UO with the FI/CO and HR functions, and presented the following to the NAOC on August 18, 1999. The following is a summary of his presentation:

Summary:

Outlined below is the introduction of SAP R/3 Financials, Human Resources/Payroll, and Logistics modules for North American business units.

Current Situation:

- *Financials*: Computron Software is used for General Ledger, Fixed Assets, Accounts Payable, and Workflow on an AS/400 computer. Additionally, Geac software is being used to run the Accounts Receivable system for several divisions on an outsourced mainframe. There are numerous interfaces that are being maintained between these systems on diverse platforms.
- *Human Resources/Payroll*: Multiple Human Resource software packages and databases are currently in use across divisions and locations. The pension and benefit programs are outsourced to Towers Perrin and the payroll is being processed on an antiquated legacy mainframe system. The systems are nonintegrated solutions that are inefficient due to fragmentation, duplication, and division-level customization. The current architecture is not suitable to meet future Bosch business requirements.
- *Logistics*: The logistics systems architecture consists of several systems loosely integrated through in-house developed software and complex interfaces. The support of this inefficient architecture is expensive and does not allow for a quick response to changing customer requirements.

Figure 3. SAP project proposal: Financial — project management organization

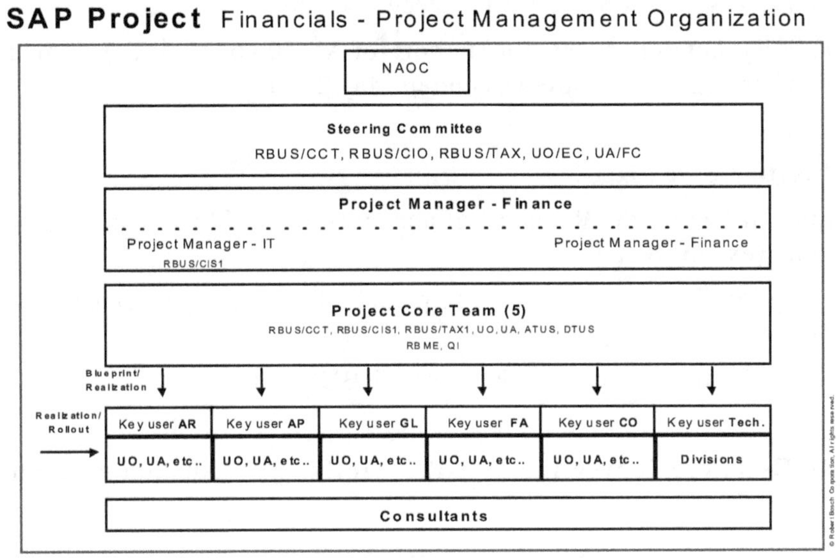

- *Objective*: The following systems would be implemented in North America:
- *Financials*: Implement the Bosch standard SAP R/3 Financial system in all RBUS divisions to provide a common solution for financial consolidations and for internal and external reporting.
- *Human Resources/Payroll*: Implement the Bosch standard SAP R/3 Human Resources/Payroll system to establish a consolidated and centralized HR and Payroll system for US Bosch locations.
- *Logistics*: Implement SAP Logistics at the UO plants.

Project Cost:

The costs for Financials and Human Resources/Payroll projects were based upon consultation with QI consultants and three external implementation partners and will be fine-tuned in the initial project phase. Logistics costs were high-level estimates to provide a total SAP cost picture — the logistics costs would again be reviewed with the NAOC prior to implementation.

Justification:

- Financials: The existing RBUS Financial system must be replaced: The current software, provided by Computron Software, will not be supported on the AS/400 computer after December 2001.
- IBM has announced that the support for AS/400 operating system required by Computron software will end in May 2000.

Benefits to be realized:

- Standardize on the RB chart of accounts across North America business units (except SBPT)
- Provide building block for subsequent SAP logistics implementation at the plants
- Simplify financial consolidations for RBUS and divisions

Human Resources/Payroll: SAP HR/Payroll implementation is required to:

- Eliminate aging mainframe legacy system
- Provide Human Resources systems where none exist today
- Standardize HR functions across divisions
- Support RB worldwide standardization through SAP R/3

Benefits to be realized:

- Reduce dependency on suppliers May & Speh, Geac, and Computron
- Possible elimination of outsourced benefit provider, Towers Perrin
- Elimination of 3rd Party Software Solutions HRPulse, Access Data bases, spreadsheets, and support
- SAP ESS (Employee Self Service) reduction of employee transaction costs

Logistics: SAP implementation is necessary to:

- Provide for RB standardization in UO plant operations including DTUS and RBME
- With SAP Financial and Human Resources/Payroll, provide integrated, closed-loop system
- Eliminate non-integrated PRMS solution, where:

- Addition and deletion of new businesses is complex and difficult
- Existence of numerous interfaces creates a complex environment
- In-house developed systems are costly to maintain and support
- Remove the requirement to upgrade PRMS from version 8.4 to 9.0

Timeframe:

- *Financials*: The total project duration is estimated to be 3.5 years from start up, with the first division to go live in 12 months. The remaining divisions are to follow over a period of 30 months.
- *Human Resources/Payroll*: Implementation of Human Resources/Payroll will start concurrently with Financials. The total project duration is estimated to be 15 months, with first pilot site to go live in 12 months and remaining sites to be implemented in 13 to 15 months.
- *Logistics*: The project is roughly estimated to be 3.5 years in duration.

Risk analysis:

Project risk is mitigated due to:

- Use of experienced U.S. implementation partners with proven implementation successes
- Use of QI and K1 resources and experience
- Formal coordination meetings between the Financial, Human Resource/Payroll, and Logistics project managers
- Staged implementation using an initial pilot implementation to prove concept
- Standard SAP software will be used without modification
- Readiness to change Bosch processes and adopt SAP standards
- Freeze development in current HR/Payroll and Financials system

Possible increased project risk would include:

- Merger and acquisition activity during project life cycle
- Logistics activity proceeding prior to QI kernel availability
- Personnel turnover during project implementation due to SAP salary inflation

Next Steps

- Select Implementation Partner
- Staff project core teams (see Figure 3 as an example for the financial project)
- Project start-up

August 1999: Feedback from NAOC

Don and the CFO, Gary Saunders, outlined the SAP direction to the North American Operating Committee during August 1999. They consulted with the management in K1 and UO and obtained general agreement that the direction was acceptable to them. The committee wanted additional information regarding the project cost drivers, specifically how much was due to:

1. "Must do" items
2. A corporate standard
3. Need to improve processes

There was pressure to reduce the spending on the SAP systems implementation. Robert Oswald (F4C) wanted to know what the current running costs were vs. the forecast costs once SAP was implemented. He also asked for estimates of costs to continue using the Computron system.

Don explained further:

We are planning to go back with a final proposal in December. Our concept has changed significantly. UO, based upon K1's suggestion, was going to integrate the implementation of the financial and logistic modules and meant higher allocations to the smaller divisions for the SAP FI implementation. And there is an issue of how much of the UO FI implementation should be included in the corporate project. Do I implement the SAP finance system to handle UA's J.D. Edward's and ATUS's PRMS, and the PRMS implementations in other divisions? Who is going to bear the cost? Will it be too costly for the smaller divisions to bear since the larger divisions are unwilling to subsidize them?

Let me just give you an idea of the benefits of what will happen if SAP R/3 is implemented. As it will run in South Bend on a centralized server farm, we will be able to reduce the number of major data centers. There will be a small data center for UAs J.D. Edwards processing, but all SAP will run centrally at South Bend. There could be standard processes for the OEM systems.

September 1999: Revisions Recommended by Mike

Mike Bieganski, vice president of Information Systems and Services, had implemented a SAP R/3 system during 1997 to 1999 for the K1 division. This division had sales in excess of $1.5 billion during 1998, producing brakes, rotors, master cylinders, antilock braking systems, and so forth for passenger cars, light trucks, and medium trucks. The K1 unit was formed when RB purchased the AlliedSignal braking division during 1996.

AlliedSignal had software running on a mainframe for which they did not have the source code. Therefore, they were not able to add or change any of the source code to make it Y2K compliant. RB thus had to replace their system. Mike was hired in 1997 to implement the SAP project. As a result of the SAP R/3 project, data centers were consolidated, and seven modules were installed at 11 sites. A total of 1,200 users were affected. Figure 4 shows the SAP project scope at K1, and Figure 5 shows the status of the K1 SAP project. Due to the

Figure 4. SAP project scope in K1

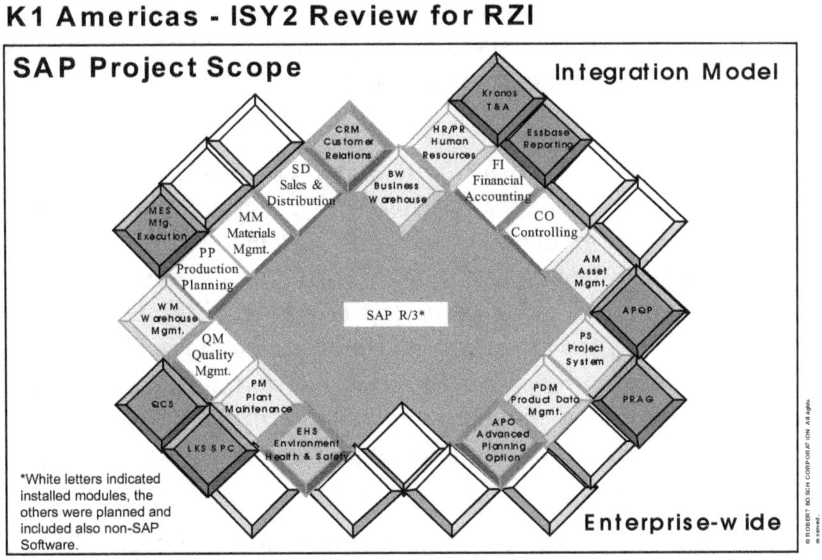

Figure 5. SAP implementation status in K1

K1 Americas - ISY2 Review for RZI

SAP Implementation Status

Seven modules installed at 11 sites
1,200 users
$13M inclusive
1,346 issues recorded
  1,029 completed
  141 duplicate/overlay/cancelled requests
  176 open items

time pressure on the implementation of this system and because of strong management support, decisions were made very quickly, and during this process Mike became an expert on implementing SAP R/3 in a large organization.

Since K1 had been asked to fund part of Don's proposal, Mike scrutinized the proposal and came up with a variety of recommendations. He believed that the implementation of SAP FI as a stand-alone system would produce few tangible benefits and would increase costs in the long run (Figure 6). He explained:

The interfaces would have to be built and then thrown away. Purchasing and receiving will not be implemented with A/P, yet vendor and related information will have to exist in multiple systems. Order processing and shipping will not be implemented with A/R, yet customer, pricing, and related information will exist in multiple systems. General Ledger will include a new chart of accounts that will not match current feeder systems; conversions will have to be thrown away if and when SAP logistics is implemented. Examples of the issues include returned shipments and debit memo processing, retro-billing and pricing issues, and mismatched payables.

Mike wanted Don to consider a new implementation approach. He suggested focusing on reducing rework by implementing an integrated suite of modules at

Figure 6. Interdependence between logistics and financials

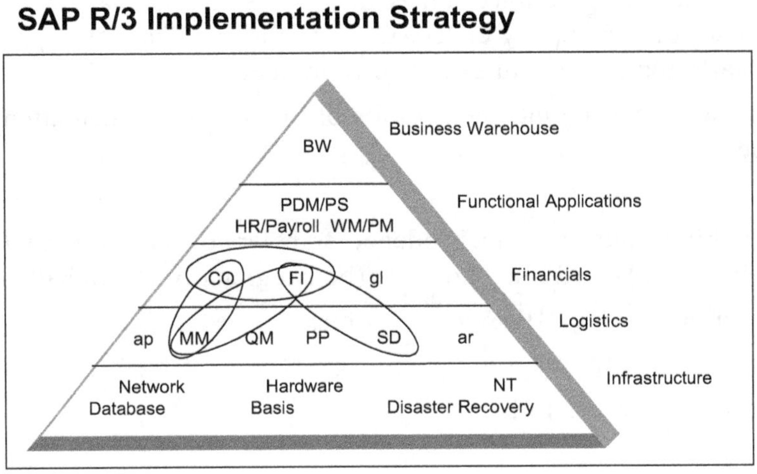

the outset, and expected a positive financial impact from this approach based upon his implementation experience. He recommended:

- Integrated K1 financials and logistics
- Integrated UO financials and logistics
- Integrated UA financials and logistics
- RBUS focus on harmonization, consolidations, mergers/acquisitions, business warehousing, and strategy for smaller business units

He expected that the new implementation approach would be standardized, and that any exceptions would be funded by the requester. He proposed avoiding interfaces and conversions and accepting the K1 implementation as the "starting-point," which could result in:

- Savings related to interfaces, conversions, and reports
- Reducing estimates for an equivalent of 14,660 consultant hours for conversions/interfaces/reporting

Other recommendations were to pilot the payroll in K1, and combine functional and technical expertise in financials. He proposed to adopt the K1 logistics model with minimum configuration changes, integrate SAP financials/logistics, and to establish separate company codes (UO/K1) for the Charleston plant product lines (Charleston manufactures the ABS product line for K1).

Mike believed that the intended results of the new implementation approach would be:

- Flexibility: business independence, transfers of plants/products between business units, and consolidations (both regionally and globally)
- Synergies among RBUS organizations

He expected the issues to be:

- Tax reporting requirements
- The SAP business warehouse (a database)
- Computron Financials: the risk in losing support by May 2000

Mike summarized his suggestions:

We are trying to figure out how do you build SAP in one big instance in the U.S. so everybody uses the same configuration based upon the QI SAP model. Or do you organize it by business units with consolidation points for reporting at RBUS? And that's what we are trying to determine. Why are we starting a system and getting bids and trying to do all these things when we don't have the very basic elements of how to harmonize? The first tasks are agreement on a specific chart of accounts, on the cost component layout, and on customer and vendor numbering. We don't have an RBUS harmonization model with RB in Germany. What I tried to tell them is not to spend any more money on infrastructure because we adopted all the Bosch infrastructure standards.

Don had concerns about Mike's recommendations and stated:

As we go through this, what Mike is suggesting is that we try to leverage K1's implementation for UO but keep them separate, as separate businesses. Well, his idea is to manage consolidations and track corporate initiatives through the use of a business warehouse. And yet, that's not what QI and RBUS have in mind. RBUS needs enough control to drive standardization into the businesses. If we're not standardizing when we do this, we will have lost a major opportunity. For example:

- *Standardization of the GL chart of accounts on RB's chart of accounts. But the mapping of that chart is very difficult for the U.S. because they do some very different things that are not common in U.S. GAAP accounting. Mike has made significant progress in analyzing this, but we must do it in a standard fashion throughout the organization.*
- *Standardizing customer numbering. When we get through with this process we need to be able to globally view our business by customer.*
- *Standardizing supplier numbering and commodity numbering. We should be able to say how much business we are doing with one of our vendors worldwide and we should understand the global requirements for a commodity and where we are currently sourcing it.*

In order to ensure this standardization and to centralize cash control, RBUS will need, at a minimum, to have a consolidation ledger, accounts payable, cash disbursement, and a fixed-asset system for external reporting. I believe that the centralization of accounts receivable is also a major opportunity for the future. Then for reporting at the North American level, there will need to be some business warehouse capability to look at total business with customers and suppliers — unless this can be done in the primary system.

Mike recommended that RBUS/K1 should pilot the HR/Payroll module at K1 in preparation for the RBUS implementation of a North American solution. Don stated:

K1 wanted a lower cost allocation for HR. We met to discuss the cost of the HR/Payroll and the CCAi Consultants participated. Mike was interested in linking HR to the logistics module and proposed running HR on the K1 SAP instance. This would not have met the requirements of RBUS to run Payroll and HR centrally and CCAi recommended against it. Based upon that discussion, we decided to run all SAP for North America on one computer; Financials, HR/Payroll, and Logistics. We believe this can be done, as the total number of users is estimated at 4,000 for all systems. But the technical and business issues will need to be reviewed.

Creation of a Revised Proposal During 1999

Don and Kamran considered the major issues that needed to be addressed in creating a revised proposal:

a. How to structure the SAP system to keep it flexible and ensure consistency
b. How to design the implementation process
c. How to justify the cost of the SAP implementation from a business standpoint

Each of these major issues will be discussed in turn.

Efforts Made by QI on Structuring and Harmonizing the SAP System

As mentioned in the section describing the R/3 strategy of Robert Bosch, there were a lot of individually designed and implemented R/3 systems that were already in place, along with systems from other vendors and self-developed systems. The integration among these systems was done by complex interfaces. Based on this experience, QI had developed a generic system architecture. The driving force to do this was to make it easier to integrate the R/3 systems all over Bosch, to leverage from multiple uses of developed solutions and thus build manageable systems. Robert Bosch was too big to work with one physical R/3 system. The reasons for the use of multiple systems were system performance, system availability, network availability, release management, flexibility of business units, and corporate culture.

Gunter Lehmann, the coordinator of the SAP Base and CCC (Customer Competence Center), was responsible for structuring and harmonizing the SAP system in QI. According to him, there are different generic possibilities for distribution of the R/3 system: distribution by function, by process, by division, by region, or by location. Simplified examples of these generic possibilities are as follows:

- *Distribution by Function*: all Bosch's HR functions are implemented using a single system
- *Distribution by Process*: all purchasing processes are implemented for Robert Bosch on a single system
- *Distribution by Division*: all the functions and data necessary for K1 are implemented on a single system

- *Distribution by Region*: all functions and data for RBUS are implemented on a single system
- *Distribution by Location*: all functions and data for South Bend are implemented on a single system

In Robert Bosch, there were some unique requirements that influenced the system architecture, namely:

- worldwide consolidation and group-wide reporting
- the maintenance and distribution of several data and settings group-wide
- divisions (GB) having specific processes/functions and data, and being responsible for them
- plants often working for different divisions and using similar functions independent of the division
- national companies (RG) acting for multiple divisions and dealing with country-specific regulations, business environments, and so forth

Lehmann created the generic system architecture shown in Figure 7 to take care of these requirements. The integration and communication between the different systems was done by the "business bus" (see Figure 8). This business bus includes different technologies such as basic communication infrastructure, middleware systems, and SAP specific technologies like ALE (Application Link Enabling, see the ALE Whitepaper issued by SAP).

To guarantee consistent accounting in each country, Lehmann assumed that standard guidelines would be followed. Examples are the chart of account, debit accounts, credit accounts, forms, reports, terms and conditions, tax regulations, and so on. This standardization effort is independent of whether the company has one centralized financial system, or a distributed but integrated financial system.

The company code in the R/3 system is defined as the smallest organizational unit for which a complete self-contained set of accounts can be drawn up for purposes of external reporting. For Bosch, the company code is defined as the accounting area. QI has defined this code as consisting of three digits followed by zero. If a company feels a need for several company codes, the company code number is expandable in its fourth digit. Details of the SAP terms are available on the Online Documentation CD. QI also developed an internal glossary for structure modeling in RB.

SAP states that to avoid unnecessary changes in the organizational structure of an R/3 system, it is appropriate to develop an organization concept for the entire

Figure 7. Generic R/3 system architecture for RB

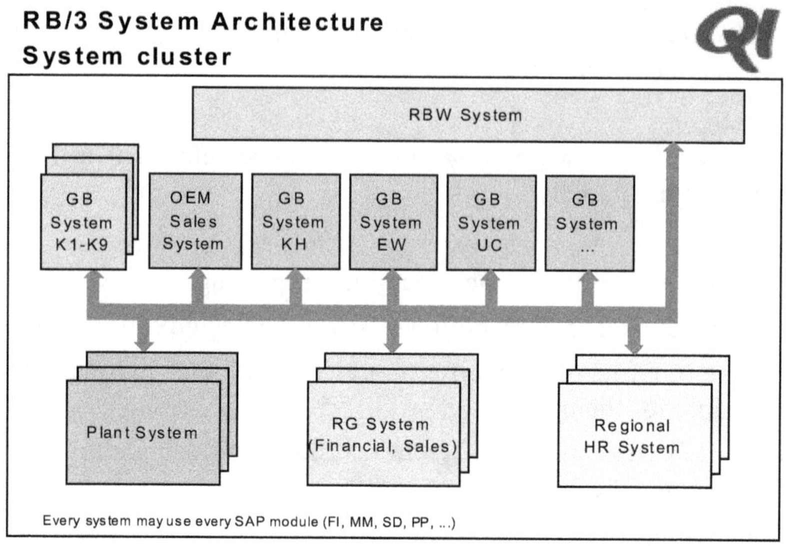

Figure 8. R/3 system architecture for RB — Business bus

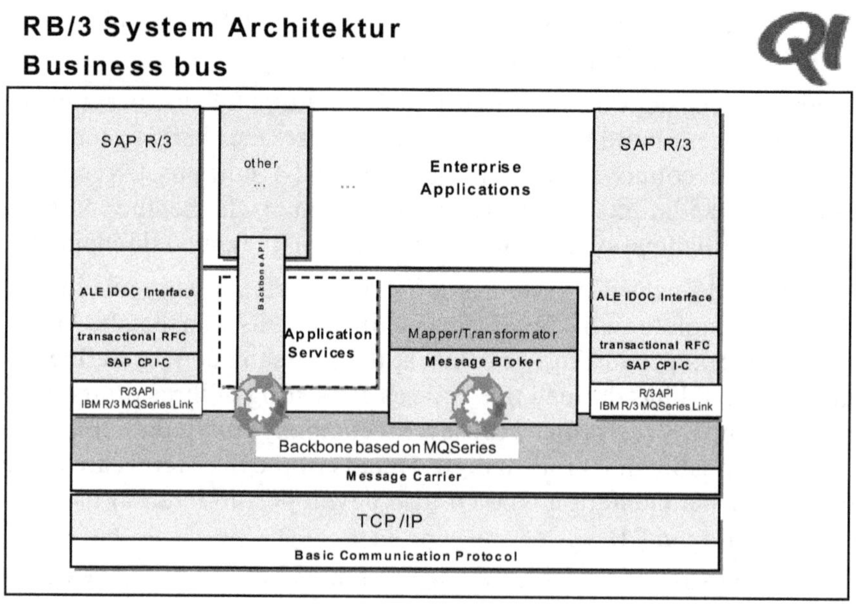

Figure 9. Enterprise modeling — the enterprise structure in R/3 (SAP AG, 1999)

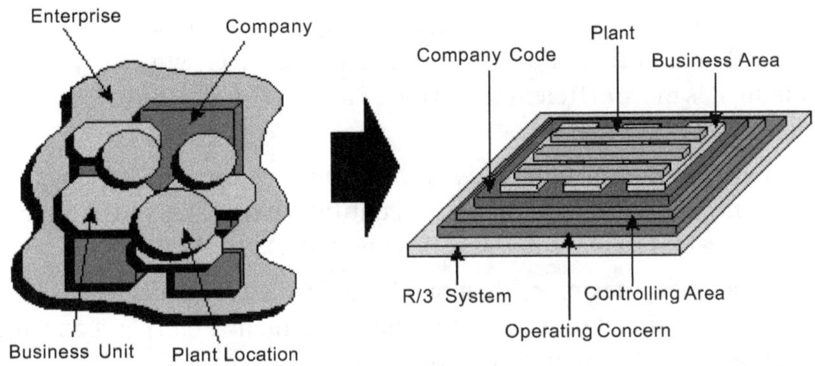

R/3 system before configuring the different application components. This means that it is useful to draw a whole picture of the application scope in order to reflect the consequences of the requirements (see Figure 9).

To ensure, for example, that one customer has only one customer number within an entity, it is necessary to define uniform organizational procedures for the Bosch enterprises worldwide. In order to facilitate the process of implementing the R/3 system across the different divisions, QI has developed standards for:

- organizational units (see the example for client and company code)
- the internal Bosch chart of accounts
- automatic account-finding in the R/3 system
- balance sheets
- structure of customer numbers
- structure of vendor numbers
- structure of material numbers
- conditions of payment
- payment terms
- groups of goods

Lehmann proposed the following structure to illustrate how to decide about the distribution of the financials functionality. The following cases are possible:

1. If R/3 is used only for financials and cost-center controlling in a country, it is most efficient to use a single centralized system (Rule 1 in the following table).
2. If an R/3 logistic system is in place, it is inevitable that big portions of FI and CO will be included. For national companies (RG) with only one logistic system, it is more efficient to run the FI and CO on this logistic system (Rule 2 in the following table).
3. If a logistic system is already in place in a country, it is possible to run different accounting systems (legal entities) with different company codes on this system (Rule 2 in the following table).
4. If a legal entity has more than one logistic system in place, it is necessary to split centralized (e.g., balance sheet, payments, etc.) and decentralized functions for the operative processes, especially CO. In this case it is necessary to interconnect the systems by interfaces (e.g., batch-input or ALE; Rule 3 in the following table).

Decision table	**RULE 1**	**RULE 2**	**RULE 3**
R/3 only for financials in a country	YES	-	-
Only one R/3 logistic system is in place in a country	-	YES	-
More than one R/3 logistic system is in place in a country	-	-	YES
One centralized R/3 system for financials	USE	-	-
Financials run the R/3 logistic system with one or more companies	-	USE	-
Multiple R/3 logistic systems with decentralized FI functions and one centralized R/3 system with centralized FI functions	-	-	USE

To ensure that different systems can cooperate easily with each other it is necessary to have guidelines for standards. Within Bosch, Lehmann designed guidelines covering, for example, the structure and enterprise modeling and the customization of an R/3 system. Enterprise modeling refers to the planning and representation of the basic structure of R/3 across all system modules, taking the strategic business processes, information flows, distribution scenarios, and R/3 business objects into account. An enterprise model represents the structure of an enterprise using the organizational units implemented in R/3, and serves to evaluate the usability of an R/3 system that has been configured in accordance with the business requirements of the enterprise. Figure 10 shows this approach in a simplified way.

For example, the client in SAP is defined as a system in which applications are integrated into a common database. In SAP terms, a logical system is a client in a database. The company's objective is to represent Bosch as a single client

Figure 10. Impacts on enterprise modeling (SAP AG, 1999)

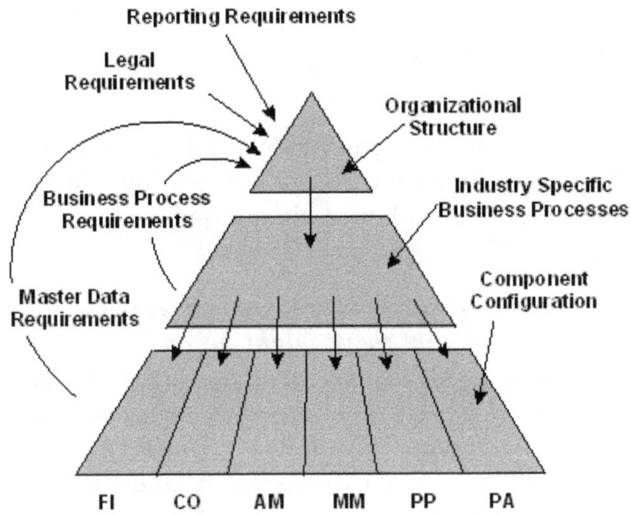

irrespective of the different divisions and locations throughout the world. This client can be distributed onto different systems. Standardized structures are therefore essential, as it is vital for messages to be easily exchanged between logical systems.

Structuring Information Systems in North America

The SAP R/3 implementations of K1 and SBPT in North America were different from the practice and guidelines of QI, largely due to the speed with which these implementations were performed. As a result, the chart of accounts used by K1 was different from that of other divisions, leading to difficulties in obtaining summary reports on performance. In the future, K1 plans to harmonize its system according to the new guidelines wherever possible. Mike stated:

Never do we want to hear that in this merger or acquisition or divestiture, the big problem is systems. And we actually have a vision in the way we put this in that if somebody said today we wanted to add a new plant, we literally could have everything up and running in a week. Now it may be that we have to write programs to interface their data into ours and to talk about business processes and go through a lot of training. But to set the parameters, put a new plant code in, very easy — perhaps in one day. We're not there yet because we have a lot of harmonizations to go through. But it's important to know that we're flexible and the business doesn't have to wait on systems to make decisions.

Don was responsible for proposing the models that would be used for structuring the planned R/3 solution, including FI/CO, HR, and logistics, in North America (see Appendices 1a-1e). He explained his reasoning as follows:

My first understanding of structuring our system was to have different instances/systems; for example there might be an HR instance, there might be an FI instance, and separate instances for logistics. UO and K1 might have separate instances or, as their business processes are similar, there might be one OEM automotive instance. UA could be a separate instance, or you could look at UA and SBPT as similar businesses with manufacturing and distribution currently distributes some SBPT products — and have a "distribution instance" for both businesses.

Another possibility is for K1 to be aligned as a GB and have one instance with the K1 businesses in France and Germany. This is again not a systems issue but a question of how Bosch wants to run the business.

After the meeting with Mike Bieganski at K1, Don was planning to run the whole SAP R/3 solution based on a single instance:

If we go back to my original model there are some issues. We would have to integrate the separate instances through ALE or batch interfaces. As I understand SAP, it is extremely difficult to write interfaces to the table structure of the data. A purchase order is no longer in a PO file, but can be spread across a large number of tables. Consultants have said that the simple batch interface that you could write in a week now takes six weeks. So, the interfaces between separate instances were adding a lot of cost to the project.

Currently, therefore, we are looking at one instance. This avoids the ALE interfaces between instances and will only require batch interfaces from the non-SAP systems. These are already standardized to Computron and we think that by using these existing interfaces with minor changes we can feed them into SAP. This would avoid significant cost and simplify the overall project.

But there are issues. With UO and K1 implementing integrated FI and logistic systems, and with corporate still controlling the overall FI picture, who does the configuration? How do you standardize, and who's responsible for what? Kamran's group should be responsible for configuring finance, but there may be some pieces of finance that relate more to product cost, and that should be down at the UO/K1 level. So, these are the issues that we are struggling with, but the major issue is that we've got to replace the current finance system, and this does not address this issue for the smaller divisions. Do we run centralized SAP Financials for all the non-SAP divisions, or do we allow them to decentralize the financial systems to J.D. Edwards, PRMS, and so forth, and only run a consolidation G/L, payables disbursements and a fixed-asset system for external reporting at corporate?

One instance implies that we are going to need large servers to service about 4,000 users. We believe the architecture can handle this workload, but this needs to be tested. Interfaces are where systems break, and having a single instance will avoid a lot of interfaces. It will be the most flexible structure allowing different views of the data, which will satisfy the complexity in reporting.

The major drawback to this approach is the issue of upgrading the software. Changing a multibillion-dollar company with numerous plants to a new release of the software at one time will be difficult. But already we have half of this company on SAP, and Mike's taking K1 to a new release now. So it's a problem of scale.

Design of the Implementation Process

A SAP consultant once commented about implementing SAP:

SAP has the flexibility of wet cement. As you're putting it in you can mold it any way you want it, but as soon as it hardens don't try and change it, because it is very difficult to change.

The experiences of implementing SAP/R3 in K1 and the basic architecture of R/3 showed Don that it was necessary to first of all go through an in-depth analysis of the whole application landscape to do the enterprise modeling (see Figure 10). These decisions had to be done carefully to meet the current business needs and to be flexible enough for changes in the business environment.

In the August NAOC meeting, the implementation plan encompassing phase 1 for the main businesses and phase 2 for the smaller businesses was presented. Krish had the following ideas about the implementation:

We start the SAP implementation study during December 1999 and January 2000. We start with an improvement phase that will tie in the financials and logistics implementation of UO with financial implementation of RBUS. Then we will go into a realization phase.

We will take the best practices of our K1 organization that has finished the implementation of SAP and come up with the delta, because there is a delta between our business processes in UO and the business processes in K1. We hope to finish the realization phase between July and December of 2000. And in December of 2000, we will start our implementation phase with two of our major plants, which are Charleston and Anderson. Once we have finished those two plants we will have our organization trained, we will have a handle on the implementation, we will understand SAP more from the IT side as well as from the user side, because that is very critical again. Then we will implement SAP in the two smaller plants, which are Albion, Indiana, and Juarez, Mexico. Our total completion date for at least the four plants, the core four plants of UO, is December of 2001. Then I think we will have 1,300 users. To do this challenging project the team members need a good understanding of R/3, and right now there are probably two people who have got an understanding of R/3 in the IT organization of UO.

There was a lot of work involved in bringing all this together and it would require a major change in the organization. Don assessed the situation:

We have all read in the Wall Street Journal about the large project failures in implementing SAP. SAP R/3 is a data-hungry system. It is conceivable that without user backing, the implementation will flounder — the users need to be involved and establish the process requirements. And overall, the business strategy needs to drive the system's direction. We need to ensure that the users and management appreciate the complexity of SAP's structuring and its future implications. It can have substantial impact, particularly if we don't do it right.

Kamran Ashraf also commented on the investment in large IT systems such as the ERP system, suggesting it had to be handled like a venture capital investment (Austin & Nolan, 1998):

If I am going to invest the money, I would like to know what I'm getting for it and I would like to know what is the risk associated with it. I mean, this is just like another venture capital investment.

Taking into account these issues, there was a lot of persuasion necessary to convince the management to support implementing SAP in the RB plants. QI released a set of guidelines, described below, in order to justify SAP R/3 implementation.

Arguments for a Business Justification for SAP R/3

Mr. Schlaefer, project manager for the huge RB/3W project, gave the following arguments in a presentation to the board members (Messrs, Hoffmann, & Dais).

SAP R/3 is Essential for Bosch Because It:

- minimizes costs
- helps customer orientation
- increases the company's flexibility on all levels
- standardizes procedures across the company
- implies Bosch's orientation on the future — the first step to new technologies
- increases communication
- increases globalization
- supports a much better/faster cooperation between customer and supplier

Cost Reduction of Process Due to SAP R/3

- SAP implementation projects are organization projects but not IT projects. Implementation of SAP could lead to reassessing existing processes, adjusting, and optimizing them.
- SAP allows better communication between the departments. Integrated SAP standard processes lead to integrated operative processes. Users are process-oriented and not task-centric. Each user is dependent on the

results of the preceding process step. This results in promptness, improved quality, and flexibility.

- Old "nice to have" processes will be eliminated.
- Profitability analysis across the company will be easier due to harmonized data.
- All applications have the same user interface, so after certain training sessions a proxy will be easier.
- Reduction in the number of experts needed for system maintenance due to the homogeneous software and reduced number of interfaces. The associated reduction in services and maintenance of legacy systems will largely outweigh the project costs.
- New developments will be purchased, not done within the company.
- There will be a unique, overall relevant price for users, which can be easily matched against those quoted by competitors.
- Estimates of the SAP R/3 user costs for an implemented system after one year are $250 per user, including the following percentages for the different categories:

<table>
<tr><td>Hardware (investments, depreciation, licenses)</td><td rowspan="3">50 %</td></tr>
<tr><td>Hardware support</td></tr>
<tr><td>Network costs</td></tr>
<tr><td>Application support maintenance</td><td rowspan="2">50 %</td></tr>
<tr><td>Application support for small/middle sized projects</td></tr>
<tr><td>Total cost per user and month (max.)</td><td>$250</td></tr>
</table>

Customer Orientation

Customer requirements are often difficult to grasp in today's quickly changing communication environment. The latest version of a technology is always the first step toward new possibilities. Using a widely spread software like SAP always carries the advantage that not every requirement of Bosch has to be developed. Accordingly, company requirements can be fulfilled more quickly. This development is comparable with the changes in product policy in the automobile industry.

Flexibility on the Company Level

In a global company, prompt actions and reactions at the company level are essential. A unique company standard, as well as innovative software, enables managers to merge, separate or reconfigure businesses or companies more rapidly and effectively.

Standards

- at the process level (i.e., a uniform calculation scheme)
- at the structure level (i.e., client, company code, plant, sales organization, which enables cross-client consolidation and reporting)
- at the level of units of measure (i.e., kilo, liter, DM, package, piece....)
- uniform codes (i.e., material number, customer number, vendor number)
- uniform material type definition
- uniform cost center structure
- uniform reporting of data for all Bosch companies' relevant chart of accounts
- differentiated and clearly defined number ranges in all areas

Orientation on the Future

It is not necessarily the best software that becomes the standard, but the software that is sold the most. Global standards, a global service, know-how and add-on net value is created as a result. Coherence and interrelation guarantee the option to participate in the next innovation step. A platform for all following developments will be created (i.e., data warehouse, strategic enterprise management, e-commerce). This point is an essential tool for motivating the employees. Many employees wish to become more qualified and thus to extend their current value.

Enhanced Communication

Each staff member is a vital part in an integrated process. The employees depend on each other more than they did before SAP. Accordingly, the communication between and amongst departments functions better. Working jointly to develop the new processes and procedures increases the cooperation between the departments and QI (user-driven project).

Increased Globalization

As a global player, Bosch will bundle its international activities within the next few years. This requires a uniform and prompt information structure in order to steer the company's global aims effectively. Standards in IT will be mandatory in order to combine all needed data quickly. An integrated SAP project should be the beginning of this process.

Better Cooperation with Customer and Supplier

Is it the customer that dictates the supplier's software? The relationship between automobile manufacturers and their suppliers will inevitably continue to grow closer. The system-based communication requires a high degree of compatibility (see Figure 11).

Mike discussed the advantages of implementing SAP R/3 in the K1 division:

Let me tell you what the difference is in our business today vs. a year ago, before SAP. I went to our St. Joseph, Michigan plant and I was talking with the cost accounting person and I asked her about SAP. She had it now for 9 months and she said, 'I hate it.' And I said, why is that? Is it because we used to have a nice way to put a bill of material in and now you have to go through four or six screens, so the navigation isn't quite there and SAP is working on it? She said, oh no, that doesn't bother me. I hate it because this system is so totally integrated that I have to talk to my materials guy every single day to stay coordinated. I just hate that. So I said, yeah but what's the problem? So the benefit, the beauty, and the problem is that SAP is so integrated it's a push-down, pop-up. Anywhere in the business that there is a problem, it's completely exposed. If you have a problem on the receiving dock, the shipping dock, the middle of operations, the problem is exposed to other people and to management. Then action could be taken to solve the

Figure 11. Order delivery vs. "common" systems

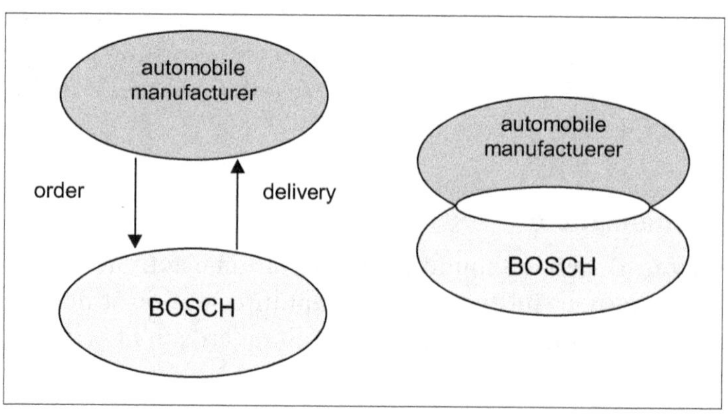

problem. So, it's just a situation where now that they have an integrated system, the same person gave me an example and said, you know, as a cost accounting person, I had my own world there. Everything was wonderful and I booked what I needed to and then at inventory time, we figured out where to match up. We got together once a year and fixed our problems. But now it is integrated I need to stay coordinated every day.

Our vice president of manufacturing heard stories of what SAP made the people do in terms of recording of data and the integration. And he wanted to see for himself and we walked him through all the details. He is a former materials person so he understood MRP very well. He walked through a lot of the details of the system and he loves it. The only thing he doesn't like is in the manufacturing world, when you make a sale, you book to the general ledger. He would like to have some buffer there. But the financial people like it because at any point in time, any day, you can fake like you are closing the books and see all of your numbers any day you want going through the month. And it's very effective.

Decision Faced by Don

Designing and implementing SAP in RBUS was a difficult change for the organization. To make this project successful, Don not only had to decide on structuring, but also had to design the implementation process. In addition, it was necessary to justify the cost needed to perform SAP implementation from a business point of view. Kamran and Don started planning their strategy to accomplish this.

Chapter Summary

Don Chauncey, vice president of IS for Robert Bosch Corporation (RBUS) had made his presentation on the direction of an SAP implementation to the RBUS North American Operating Committee (NAOC) during August 1999. The plan was to introduce SAP R/3 Financials, Human Resources/Payroll, and Logistics modules for RBUS and UO (North American Automotive Original Equipment). He had complied with QI's (Querschnittsbereich Infomationsverarbeitung — Division for Information Technology of Robert Bosch GmbH) request to define a project for the implementation of SAP in the North American Bosch businesses. Before proceeding further, the NAOC requested additional information regarding the project costs, specifically how much was due to: (1) "have to do" items, (2) a corporate standard, and (3) the need to improve processes. The NAOC also wanted to know what the running costs were vs. the forecasted costs when SAP would be implemented. The critical concern of the NAOC was the overall cost of the implementation of the SAP system.

This was a major challenge for Don as he started working on the proposal revision. In Germany, QI established SAP as the standard software to be used within Robert Bosch. Mike Bieganski, VP of Information Systems and Services at the K1 Division, presented possible alternatives that could be adapted to address the concerns of NAOC. Krish Kumar, director of Corporate Information Technology for UO, was certain that implementation of SAP R/3 in his division was imperative. Kamran Ashraf and Don started planning the strategy to address the major issues:

- how they should structure the SAP system to keep it flexible and ensure consistency.
- how they should design the implementation process how could they justify the cost of the SAP implementation from a business standpoint.

Team Assignments and Discussion Questions

1. For the Bosch group as a whole, different strategic approaches might be valuable. Assume you are an assistant to Dr. Eggensperger, who has asked you to analyze the following approaches to implementation:

   (a) A standardized ERP approach for the entire Bosch group
   (b) A domestic ERP approach for each country
   (c) A business-unit-based ERP approach within the global operations

2. Reading assignments with presentation:

   (a) Describe the Application Link Enabling (ALE) technology incorporated in SAP and show how it might be applicable to RBUS.

(b) Describe the Enterprise Modelling Methodology of SAP and show why it is important for a company such as RBUS.

3. Develop suggestions for standardization with respect to the need for harmonization and flexibility in a global company.
4. Analyze impacts on structuring the R/3 system in an organization. Describe possible alternatives to the presented structure of the SAP R/3 system for Robert Bosch Corporation.
5. Propose an approach for implementing the SAP R/3 system for Robert Bosch Corporation.
6. Assume Don Chauncey's role to make a new proposal: What would be the best way to sell the technical structure and implementation plan to NAOC?

Small Group Discussion

7. Assume the role of Mr. Lehmann. What is the impact of structuring the R/3 system in an organization? Explain the alternatives to the presented structure for RBUS. What architecture will you recommend for RBUS?
8. Assume the role of Mrs. Bauer. Which architecture should be adopted by RB GmbH and RBUS? Why?
9. Assume the role of Don Chauncey. What structure should be used by RBUS? Why?

Large Group Discussion

10. Given that the three teams represented by Mrs. Bauer, Mr. Lehmann, and Mr. Chauncey (Assignments 7, 8, and 9) are likely to recommend different IT architectures, work together as a class and come up with an IT architecture and implementation plan that could be presented to the Robert Bosch GmbH board.

References

Austin, R. D., & Nolan, R. L. (1998). *Manage ERP initiatives as new ventures, not IT projects.* Working Paper Harvard Business School (pp. 99-024). Revised December 1998.

SAP AG (1999). SAP R/3 Online Documentation 4.0. Retrieved September 10, 1999, from http://help.sap.com/saphelp_40b/helpdata/en/e1/8e51341a06084de10000009b38f83b/frameset.htm

Appendix 1

Appendix 1a. Alternatives for structuring the SAP solution

RBUS/UO/K1 Meeting - November 4, 1998
SAP Direction

OEM Implementation Alternatives:

Centralized financial, human resource and manufacturing systems.

Centralized financial and human resource systems with decentralized manufacturing systems.

Centralized human resource system with decentralized financial and manufacturing systems

Centralized human resource system with decentralized financial and manufacturing systems based upon the same SAP customization.

Centralized human resource system with decentralized financial and manufacturing systems with different SAP customization.

Appendix 1b. Alternative one for structuring the SAP solution

RBUS/UO/K1 Meeting - November 4, 1998
SAP Alternative One

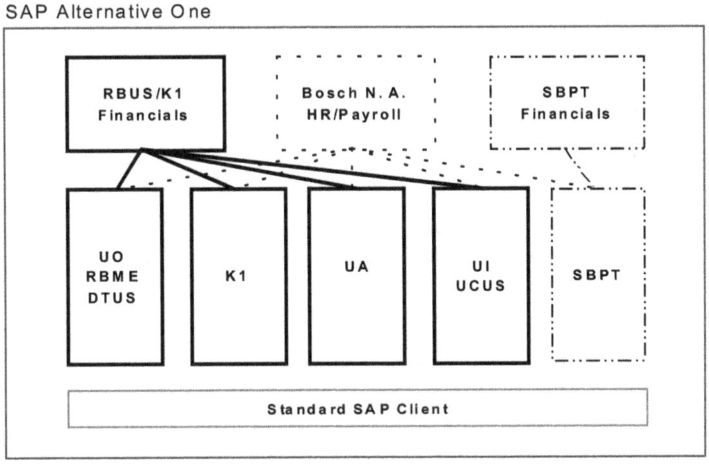

Appendix 1c. Alternative two for structuring the SAP solution

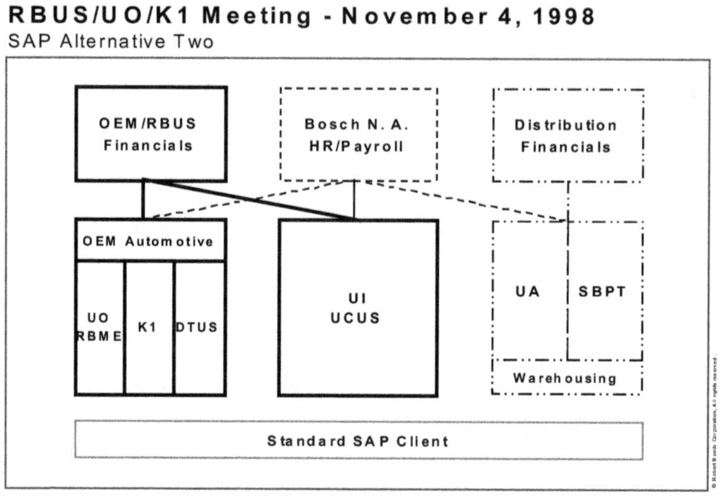

Appendix 1d. Alternative three for structuring the SAP solution

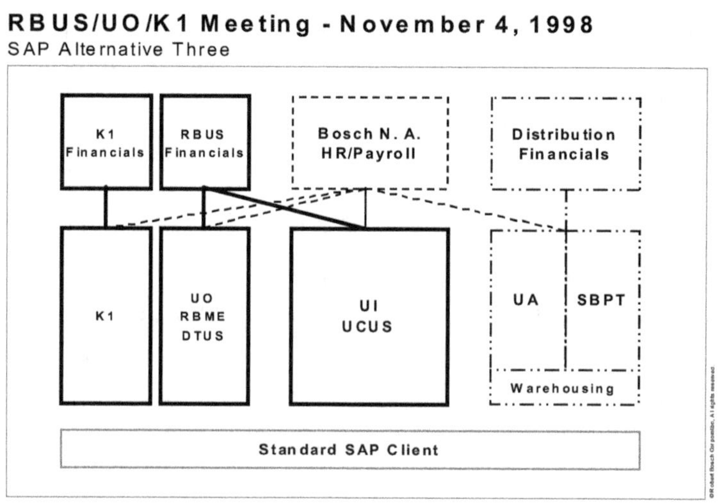

Appendix 1e. Alternative four for structuring the SAP solution with the same and different customization

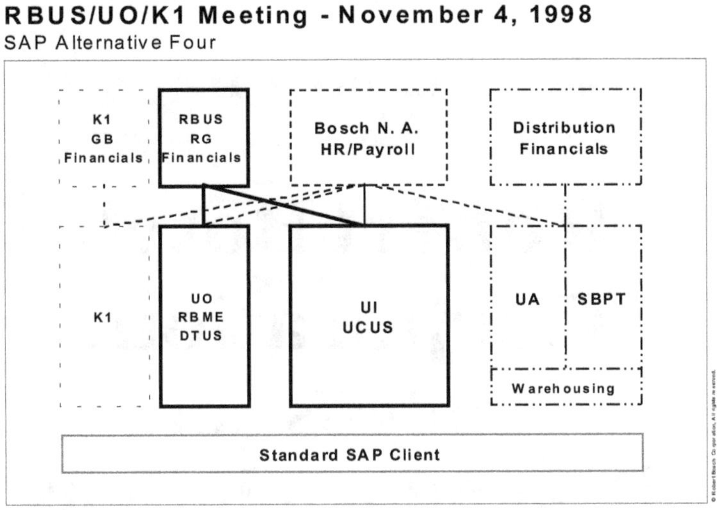

Chapter VIII

Robert Bosch Corporation: Analysis and Recommendations

Learning Outcomes

- Analyze the case study presented in earlier chapters (1992-1999), noting the connections between theory and practice
- Compare your analysis with that of the authors
- Generalize from the particular example and come up with suggestions for managers implementing new ERP systems

Overview

In the Robert Bosch case study, we presented the strategic issues facing the CIOs of the information technology division of the corporation, Robert Bosch GmbH, and its U.S. subsidiary, Robert Bosch US (RBUS). Bosch had traditionally followed an international/multidomestic strategy, using multiple information systems for its plants, divisions, and business sectors. Due to increased pressure to cut costs, increase interchangeability of products among the many plants worldwide, and fulfill customer requirements, the corporation was moving to a global strategy. In this chapter, we provide our analysis of the Robert Bosch case study and answer the questions that were raised at the ends of Chapters 6 and 7. The analysis is designed to generate debate and discussion among students, thereby adding to their learning about SAP R/3 implementation. It does not in any way represent the position and opinion of Robert Bosch GmbH and Robert Bosch U.S. In Chapter 9, we will discuss the strategies and actions actually undertaken by the IT divisions of Robert Bosch GmbH and Robert Bosch US during the implementation period, 2001-2004.

Analysis and Recommendations

Analysis of Strategies Used to Integrate Multiple Information Systems at Robert Bosch Corporation (1992-1999)

We will analyze the case study presented in Chapter 6 in this section. The questions are mainly concerned with organizational issues and the need for RB to use a standardized IT system throughout its organization. The case provides financial data in order to emphasize that the profit level was not satisfactory in 1998, and the company needed to change its focus on developing systems based on the divisional model.

Organizational Structure of Robert Bosch GmbH and RBUS

Robert Bosch is a good example of how an organization evolves over time through acquisitions and mergers, and is a practical example of the organizational theory and change management issues discussed in Chapters II and III. RB GmbH has ventured out beyond "what it is good at." The company did not just settle down in its core automotive business and stop growing at that point, but realized that since its name was out there and it had a good customer base, it should capitalize on this and try its hand at other things. This is where the mergers, acquisitions, and joint ventures began to play a significant role in the company's plans. RB did all it could to diversify the company; and did it very well, gaining a foothold in areas ranging from communication technology to power tools and even household goods. RB's structure is constantly changing due to industry change, and that is what sustains RB as a market leader and innovator.

The organizational structure of the Robert Bosch GmbH follows the divisional model, in this case consisting of the K-divisions, communication technology

Table 1. Year-to-year changes in the income/sales and sales/associate ratios for RBUS

	1989	1990	1991	1992	1993	1994	1995	1996	1997	1998
Income/ Sales	2.5%	0.7%	-1.4%	0.7%	2.8%	5%	4.9%	2.9%	2%	2.3%
Sales/ Associate ($M)	0.22	0.23	0.23	0.27	0.32	0.36	0.34	0.32	0.31	0.29

divisions, consumer goods divisions and capital goods divisions. These divisions are accompanied by corporate departments like QI. The structure of the Board shows that the members have a division responsibility, a responsibility for subsidiaries in other countries, and the responsibility for different corporate departments. Basically, this divisional structure is valid for the whole group, but if you look at the structure in the U.S. you will find differences. For example, UO is part of both the U.S. automotive OEM division and the K1 division. From a theoretical standpoint, the division structure should be valid generally so that the leadership can be concentrated at the divisional management, allowing the different countries to act as hosts of the divisions with a mission to support the infrastructure. However, if the customer and business environment requirements are very different from country to country, it often makes sense to deviate from this principle. Alternatively, it could be a question of size. At RBUS the structure is the way it is for largely historical reasons. Originally, UO represented the automotive business. The acquisition of the braking division of Allied Signals in 1996 doubled the automotive business, so this was handled as a separate part. Another factor is that the Allied Signal business is not limited to the U.S. In effect, the integration of the recent mergers and acquisitions is still ongoing.

RBUS's sales have grown from $1.8 billion in 1990 to $5.3 billion in 1998. However, its income from operations has not increased accordingly. Table 1 shows how the ratios of income to sales and sales per associate changed over the ten-year period from 1989 to 1998.

This table shows that the income/sales ratio improved in the early part of this period, reaching a peak of around 5% in 1994-1995, but then declined. This should be of concern to management. The sales in millions of dollars per associate also increased up to 1994 and then decreased. These data seem to indicate that the acquisitions and mergers are having a major impact on the financial health of the company. It appears that the company is struggling to get its systems and new operations to work smoothly together.

For comparison, RB GmbH's sales have grown from 30.6 billion DM in 1989 to 50.3 billion DM in 1998, but the net income has grown at a slower rate. The sales per associate ratio shows a continuous increase (Table 2). However, the income to sales ratio fluctuated between 1.2% to 2.0% (except in 1997), which is a relatively low level. This confirms the statement made by Dr. Scholl that the

Table 2. Year-to-year changes in the income/sales and sales/associate ratios for Robert Bosch GmbH

	1989	1990	1991	1992	1993	1994	1995	1996	1997	1998
Income/sales	2.05%	1.76%	1.61%	1.49%	1.31%	1.49%	1.53%	1.22%	3.54%	1.69%
Sales/associate in million DM	0.17	0.18	0.19	0.19	0.20	0.22	0.23	0.24	0.26	0.27

profit level is not satisfactory. Given the need for increased profitability and the multiple information systems that were being used in the company, there was a need to align the IT structure better with the organizational structure.

Interdependence and Connection between the Organizational Environment and the IS Structure of RBUS

An interesting fact about RB GmbH is that the "shareholder" is actually a large industrial foundation and the profits are mainly used for self-financing and to pursue nonprofit and charitable goals. This is a very different structure of organization from those normally found in the American and European markets. From the case study, it emerges that the organization is willing to engage in an unusually high level of internal debate, discussions, and arguments in order to arrive at decisions. The company does not have the typical short-term "hire-fire" mentality, where their business strategy depends solely on the next quarter's results.

In the literature of strategic management, Chandler's aphorism "Structure Follows Strategy" is frequently quoted (Chandler, 1977). By analogy, we could say that for Robert Bosch "IT Strategy Follows Business Strategy." Don Chauncy states: "We were driving this, and accomplished it credibly because we understood the business process and were able to put all the acquired companies on to our systems. We are now almost entirely on AS/400 systems, including PRMS, Computron, and J.D. Edwards. The only mainframe systems remaining are the Human Resource/Payroll and the Accounts Receivable Systems running on an outsourced mainframe." This statement shows that the IT people effectively drove the integration of new businesses, rather than the business people. The justification for this situation was based on the cost reductions that could be gained by migrating from the old mainframe-based systems.

There is a clear connection between the organization's structure and the IT structure. Top-level executives ultimately approve all IS decisions and activities, so they must be involved in organizing the IT structure at the very top. Heading the IT structure is NAOC, the North American Operating Committee. This committee is composed of Robert Oswald, the head of RBUS, Gary Saunders,

Vice President and CFO of RBUS, the presidents of K1, SBPT, UA, UI, and an executive from RBME, all of whom are top management executives in Bosch. Furthermore, setting up a CIS (Corporate Information Steering) Committee was a very good management decision. Having someone with the depth of understanding of IT at a level that top-level management cannot have, and who is able to communicate company IT needs to them, so that top management is not just "signing off" on any new technology or change that their IS department comes up with, is a very well thought out strategy. The organization has someone who evaluates IS decisions and desires and comes to conclusions as to whether or not these changes or technologies would be beneficial for the company. Depending on who the steering committee members are, "politics" and "power struggles" may begin to influence the decisions that are being made, but having a steering committee is still a good idea, as it provides the vital link or connection between top management and the IS department that every organization needs. Top management at Bosch seems to care about not just "signing off" on things. They want and need to know what is going on and this structure enables them to do so.

Some inconsistencies in the IT structure reflect the organization structure of RBUS. For example, the functions of Krish Kumar, the CIO of UO, seem to be similar to those of Mike Bieganski, the CIO of K1. Each individual has his own vision of how the IT systems need to be integrated. On a corporate level, the financials and the HR functions are supported by centralized IT systems, so it is necessary to have interfaces between these corporate systems and the divisional systems in the K1 and UO plants and in other businesses such as SBPT and UA. Such a heterogeneous IT infrastructure fits the business from a corporate point of view if the businesses are more or less independent. However, if there is any degree of interdependence interfaces are necessary, which are costly. Krish stated that he works with a lot of different systems from different vendors in addition to some self-developed systems. Thus, he is not able to obtain available information in a timely fashion because the information is distributed among several different and incompatible systems. A real-time integration between systems from different sources is very difficult and often impossible. Given the importance of financial functions to the company, it is critical that these functions be integrated, so we will discuss the issues faced by Don during 1998 next.

IT Solutions Addressing the Challenges Faced By UO, RBUS Financials and RBUS HR Functions

Don Chauncey, CIO of RBUS, needs to find a solution for managing the company's financials function. This is not a business decision but technology-driven, as the support for the AS/400-based Computron solution currently used will end in December 2001. The easiest solution would be to change the hardware platform, so no further changes will be necessary. This would result in no changes for the user and only a technical change for the interfaces, but the data structure would be unaffected.

In addition, the corporate HR function needs a new system. This requirement is primarily business-driven, so it is necessary to choose the solution which best meets the business requirements. As the HR people prefer PeopleSoft and there is already some expertise with this application among the DTUS business staff, this could be a good solution for this area.

From a corporate point of view no one seems to have the detailed requirements for the UO division. When the standardization process begins, the systems have to be interconnected by interfaces. So UO has to identify their preferred solution on their own. Visiting the Web site of Computer Associates, the vendor of PRMS, reveals that they have many innovative features and the existing AS/400 platform could be preserved. Another question is to what extent UO could benefit from the experience of SAP implementation in K1.

From a corporate point of view, Don has to take into consideration what the standard in the Bosch group is. Don noted: "PRMS was the Bosch standard here as well as in other parts of the world, and the QI standards allow Bosch businesses to remain on PRMS. However, when new ERP systems become necessary to replace PRMS, the new standard will be SAP."

So the question is: How will Don deal with this directive? From an RBUS standpoint, Don has to prefer the solution which meets the local requirements best and which is acceptable from a business and economic point of view. Looking at the functionality of SAP R/3, all the requirements could be addressed but this would be a huge investment, not only in software but also in hardware and training for the business and IT people. This is not solely an IT question, and it is vital that the members of NAOC have a logical and coherent IT strategy for all the regional RB components.

Once Don deals with the financials, the next critical issue facing RBUS are the multiple systems that are used by UO. We will discuss the issues related to integrating these systems next.

IT Solutions Addressing the Challenges Faced By UO Functions

In 1998, UO used the PRMS system for manufacturing operations; an in-house developed system for shipping, billing, and invoicing; and EDI for customers and suppliers. Overall, they have 11 systems that address their total business cycle. Their EDI system is tightly integrated with the manufacturing and demand management functions. Krish Kumar described the existing system thus: "We do not have a very sound application layer. Because of nonintegrated system solutions, we have created pockets of information gaps between the functional departments, sales, engineering, manufacturing, distribution, logistics, and financials. We need to ensure that the design integrates with the other components of the car."

Any solution that is adopted by UO has to include the following:

a. A strategy to address business requirements,
b. An IT strategy to meet the growing and changing business requirements,
c. A function to provide real-time information to customers.

Krish had to decide on the best strategy in order to meet the organization's needs for information. One option was for him to adapt existing ERP systems such as SAP, Oracle, Baan, or PeopleSoft in order to address these needs. Typically, ERP systems not only cover the manufacturing functions but also the commercial functions, so the question will arise to what extent should RBUS use the selected system. As a supplier in the automotive business, UO is dependent on the automaker companies. These customers are very powerful and they will play a large role in determining future requirements, so from a strategic point of view it may be preferable to look for a software vendor who is already in the automobile business.

At this point, RBUS has many different information systems that take care of business needs due to mergers and acquisitions. These systems include:

1. *PRMS*: This is a materials management and production management system implemented in the plants. UO uses this system to support its manufacturing operations.
2. *Computron Software*: This is a system that supports general ledger, fixed assets, and accounts payable functions and is implemented at the RBUS headquarters.
3. *Payroll Software*: RBUS has outsourced this function to the company May & Speh, who use a mainframe to process the payroll application.
4. *SAP R/3*: This is a software system developed and supported by SAP Corporation. K1 in the U.S. has implemented seven modules to support eleven plants. These modules cover materials management, sales and distribution, production planning, quality management, and some of the financial accounting and controlling functions.
5. *Ten Other Systems*: UO uses a number of other systems to support their total business cycle. The demand management system determines what

they have to buy based on engineering data. They tie these systems to the engineering aspects of the organization and build the product using manufacturing methodologies. They also use a variety of shipping, invoicing, and billing systems and yet another system to build a financial summary for internal reporting, as well as for external reporting.

6. *EDI System*: This system is used to connect to customers in order to perform demand management functions. UO uses EDI systems to connect to 38 Original Equipment Manufacturers and to conduct transactions with their 135 biggest suppliers.
7. *EVA System*: This is a German system that is used to facilitate order entry for the UO division.
8. *EMS System*: This system is used to create documentation for U.S. Customs for the export of manufactured product and noninventory exports by UO.
9. *PeopleSoft*: The payroll module of PeopleSoft has been implemented in the DTUS division (Diesel Technology U.S. division) to run its payroll.
10. *J.D. Edwards*: This is an integrated ERP system used in the aftermarket division of the U.S. operation (the UA division).

Having analyzed the alignment between the organization and IT systems, and also having discussed the issues involved in integrating the financial functions and the systems used in UO, we are now ready to discuss the advantages and disadvantages of either staying with the multiple IS or migrating to a single SAP R/3 system throughout RBUS.

Analysis of the Plan to Continue With the Current Implementation of Multiple Information Systems

Advantages:

- All these systems except the HR-application have been changed in the last five years.
- The implementation of PRMS and Computron software was completed only one year ago.
- "Don't change a running system!"
- The corporate strategy on implementing the ERP system has not yet been determined.
- Reduced costs, since systems have to be changed only in minor details.
- Users do not have to be retrained, so lower training costs.

- Given RBUS's reasonable performance, it is important to control costs.
- Possible to follow "best-of-breed" principles.
- Unless a strong improvement in sales would result, the strategy to shift to SAP R/3 might be a very expensive option. Also, the IT team does not seem to be confident that the implementation will be a success.

Disadvantages:

- A valuable opportunity to integrate operations will be lost if the current implementations continue.
- Increasing costs are occurring at the plant level due to nonintegrated systems.
- RBUS's IT strategies will deviate from QI's directive to implement SAP R/3 if any new application is needed. This is likely to lead to a serious situation.
- As the experience with K1 indicates, once the SAP R/3 system is implemented, benefits will accrue. It is important for the IT division to be a leader in making these changes to the systems and procedures.
- Krish at UO indicates that he is ready to change to the SAP R/3 system and that it is important to make the migration work.
- NAOC has only questioned the plan; they have not rejected it. Therefore, Don could work with the other CIOs and revise the plan.
- Implementing the SAP R/3 system would guarantee RBUS a strong leadership role in the U.S. market.

Analysis of the Plan to Implement SAP R/3 System Throughout RBUS

Advantages:

- Robert Bosch GmbH is moving from a multidomestic to a global strategy, as indicated by the information provided in the case study. It is thus imperative to implement a standard system such as SAP R/3 throughout the organization.
- The company seems to be struggling to align its systems and make its operations work smoothly together. If the company is able to implement a system such as SAP R/3, significant cost reductions and business benefits should result.
- Implementing SAP R/3 would emphasize the role of the IT division at RBUS as a leader in change management. This is very good experience for the staff, boosting their morale and enhancing their value to the company.

- QI is committed to supporting RBUS with staff and know-how for the SAP R/3 implementation.
- Since RBUS is focused on a long-term business strategy, it would be very appropriate to invest the millions of dollars needed to implement SAP R/3 so that benefits could accrue over the course of time.
- The success of SAP R/3 in K1 indicates that the company's U.S. employees are receptive to the implementation of SAP R/3 in their plants.
- The integrated nature of operations would make it possible for RBUS to continue to be a strong leader in the market.
- The new system could address future business growth and the increasing complexity of operations.

Major issues in transferring the multiple information systems to the SAP R/3 System:

Several major issues arise, namely the requirement for a new HR system, the imminent withdrawal of vendor support for Computron software, incompatibility issues affecting the current systems, the approach needed to construct the SAP R/3 system as a single huge instance, and cost issues.

a. *Requirements for a New HR System*: The payroll system is based on an old mainframe that no longer fulfills the company's HR requirements adequately.

b. *Withdrawing of Vendor Support*: Computron has notified RBUS that they plan to withdraw support for their current financial system by the end of the year 2001.

c. *Incompatibility between Current Systems*: There are pockets of implementation gaps between the functional departments, sales, engineering, manufacturing, distribution, logistics, and financials, leading to a failure to address business growth and the increasing complexity of operations. For example, the existing systems are unable to integrate new designs with the other components of the car.

d. *Venture Capital Mentality*: Kamran recommended that the SAP R/3 implementation be managed as a venture capital investment, and not as a simple IT project.

e. *Cost Issues*: QI seemed to indicate to RBUS that "it is efficiency that is essential for QI, not the total costs nor the number of employees." In contrast, Mr. Oswald questioned Don on the project cost drivers, in particular how much would be budgeted for "must do" items, corporate standards, and improving processes. He also wanted to know what the current running costs were vs. the forecast costs once SAP was implemented. K1 was expected to fund part of the proposal and asked many difficult questions about the implementation strategy.

Overall, this analysis shows that it would be more beneficial for RBUS to migrate to the SAP R/3 system. Given this choice, we will now analyze how RBUS went about migrating to the SAP R/3 system, as discussed in Chapter VII.

Analysis of Difficulties in Implementing SAP R/3 at RB GmbH and RBUS During 1992-1999

Why Implement SAP R/3 Company-Wide?: The Need for Standardization, Harmonization, and Flexibility in a Global Company

The first questions facing RBUS are: why does SAP R/3 need to be implemented company-wide, and why does the system have to be similar to those used in other regions around the world? Krish Kumar stressed the need for a global outlook when discussing fuel injector production in Charleston and in Bamberg.

Let us take the example of the fuel injector produced in Charleston, which is the same as one produced by our Bamberg, Germany plant. If the product planner in Charleston, South Carolina, can sign on and see the same applications in the two plants, we can increase his effectiveness tremendously. He might notice the fluctuations in demand. He might notice that the Charleston plant lacked capacity to address the demand here, and schedule the line two days later. Perhaps this product is being produced this afternoon in Bamberg, which can therefore become a second-tier supplier. What has he done in this process? He has reduced the inventory carrying cost across the supply chain. In addition, he has scheduled and optimized not only our delivery to the customer but also the delivery of the product from Bamberg to Charleston. A prerequisite to make this happen is the alignment of the processes between the different Bosch plants.

With regard to the standardization issue, Don Chauncey also considered it vital:

RBUS needs enough control to drive standardization into the businesses. If we're not standardizing when we do this, we will have lost a major opportunity. For example:

- *Standardization of GL chart of accounts on RB's chart of accounts. But the mapping of that chart is very difficult for the U.S. because they do some very different things that are not common in U.S.*
- *GAAP Accounting. Mike has made significant progress in analyzing this, but we must do it in a standard fashion throughout the organization.*
- *Standardizing customer numbering. When we get through with this process we need to be able to globally view our business by customer.*
- *Standardizing supplier numbering and commodity numbering. We should be able to say how much business we are doing with one of our vendors worldwide and we should understand the global requirements for a commodity and where we are currently sourcing it.*

These statements and the analysis performed earlier in this chapter show that it is critical for RB GmbH and RBUS to use SAP R/3 throughout the company. This will standardize the data elements, harmonize the business process, and provide more flexibility in operating the company globally.

What are the Issues in Standardizing Operations at RB?: Organizational Level and Scope

If RBUS wants to standardize in the context of the R/3 implementation, some basic decisions have to be made. There are two dimensions. The first is the *organizational level* of standardization, in this case, whether to standardize on a regional level, a division level, or a company level. The second is the standardization *scope*, namely whether to standardize according to the structure of the customer number or the content. Ford, for example, has been assigned the number 123. A further decision has to be made about the objects that must be included in this standardization process. Don provides an example of the charts of account, the customer, and vendor number. In general these are all master data. A further category is the control data that influences the control of the data processing by the system, such as document type, posting key, or material type. In terms of the R/3 context, this control data is fixed during the customizing process. Standardized content of the master data enables easy reporting or availability checks without the need for mapping tables. Standardized control data allows easy roll-outs of pre-customized systems.

The kind of standardization adopted has different implications for the company's operations. For example, if there is a standard within the division for all regions, it is necessary to have an agreement and a well-defined procedure. This does not avoid differences between divisions, however, so it may not be possible to identify the sales to one particular customer across the different divisions. If a regional standard is chosen across the divisions, for example for the whole of North America, the coordination effort is limited to the country level and it is still not possible to identify the global sales to a single customer. If there is a need for a worldwide view across the divisions, it is necessary to implement a central point of standardization, for customer numbering, for example. Which standardization level will be most helpful depends on the type of information demanded by the company's decision-makers. The type of standardization has direct implications for the implementation of the corresponding business processes. In order to define a unique number for each vendor, it is necessary to assign the responsibility for vendor numbering to a central office. The organizational solution also has to meet the specified requirements for the reaction time. For example, it makes little sense to utilize a bureaucratic and time-consuming process for vendor numbering, requiring users to wait a week to record a new order for a new vendor. Therefore, RBUS has to devise a scheme whereby harmonization and standardization can be done quickly. It is possible to create a Web-based system for such an implementation of the standards. If the SAP R/3 process can be connected to this Web database of standards, then new numbers can be created for customers instantly and these can then be consistently applied across all the company's divisions.

Once the elements are standardized across the company, issues about structuring the information systems at RB arise.

What are the Organizational Issues Involved in Structuring SAP R/3 Systems at RB?

The structuring of the R/3 system covers two areas:

a. the organizational structure of the R/3 system (enterprise modeling)
b. the system topology (physical distribution of applications to different systems)

Let's first discuss the organizational issues. The organizational structure has to meet the normal requirements for resource-efficient business processes and adequate internal and external reporting. The following example will explain the consequences if the SAP R/3 system is structured in different ways. From the reporting point of view, it is necessary that each legal entity is represented in the R/3 system by at least one company code, although it is possible to have more than one company code for a legal entity. This is not a problem when aggregate reports are produced, but it is then impossible to separate afterwards. A real factory could be represented using multiple company codes and the SAP R/3 system could recognize them as the same, but if the plant's code has been combined with other factories in the same RBUS division with a common company code, then it is impossible to represent this factory as a separate enterprise. This example shows that it is preferable to identify the legal entities separately in the RBUS situation. It is important to make a company policy so that as RBUS merges with new units, the codes created to represent the new units are unique. These requirements are mainly driven by the company's business strategy. It is also critical to understand the enterprise modeling methodology used by SAP, and we will discuss this next.

What is the Enterprise Modeling Methodology of SAP?

Mapping and modeling the organizational structure of an enterprise using the organizational units feature in R/3 is one of the initial steps involved in configuring an R/3 system. A global module approach promises greater efficiency and reliability for this purpose.

Enterprise modeling refers to the planning and representation of the basic structure of R/3 across all the system modules by taking into account the strategic business processes, information flows, distribution scenarios, and R/3 business objects. An enterprise model represents the structure of an enterprise using the organizational units feature implemented in R/3. It serves to evaluate the usability of an R/3 system that has been configured in accordance with the business requirements of an enterprise. The following questions should be answered when these enterprise models are evaluated:

- Is all the strategic information required for strategic decisions or for the daily operation of the enterprise available in a suitable form?
- Are all core business processes supported?
- Is it possible for all of the R/3 business objects (sales orders, requisitions, and so on) to be used as required?
- Can the organizational structure be distributed to several cooperating R/3 systems?

In a structure scenario, the legal entities, areas, business locations, and so forth of an enterprise are represented by the organizational units implemented in R/3 (for example, company codes, business areas, plants, and so on). The overall structure of the enterprise is represented by the way in which these organizational units are linked and assigned.

Figure 1. Structure scenarios

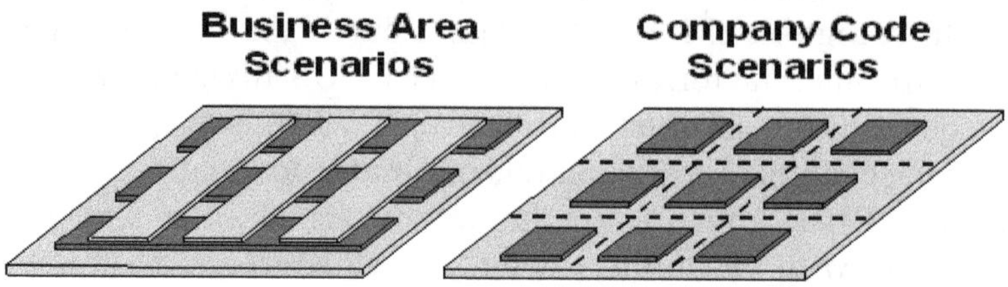

For example, as shown in Figure 1, structural scenarios can be divided into:

- Business Area Scenarios
- Company Code Scenarios

These two types of scenarios differ with regard to the way in which strategic business units are represented in R/3. In business area scenarios, strategic business units are represented by business areas. In company code scenarios, strategic business units are represented by company codes. In the second case, for example, it is relatively easy to migrate a real business unit from one country to another: To move it in the system only the aggregation level must be changed. In contrast, the business area scenario makes it less complex to handle the operational business, but it involves a lot of effort if organizational changes are required.

Another factor that determines how the organizational structure is defined is based on the R/3 system topology. The various structure scenarios, therefore, are described within the context of different system topology models:

- *Cross-Enterprise*: one central R/3 system
- *Regional-Oriented System Topology*: the business area scenarios
- *Business Unit-Oriented System Topologies*: the company code scenarios

These system topologies and structure scenarios are often blended in practice. Regional-oriented accounting systems, for example, are often combined with business unit-oriented logistics systems. The organizational structure of an enterprise is influenced by various factors, as shown in Figure 2.

The internal divisions of an enterprise can be based on the markets to which it supplies its products, or on the business functions of the enterprise. Management

Figure 2. Factors that influence organization structure

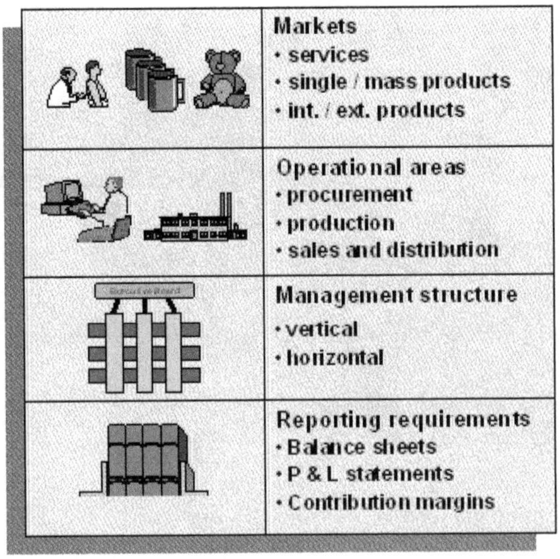

structures, for example, can be hierarchical, matrix-based, or set up as a network of independent projects. Typical reporting requirements include balance sheets, profit and loss statements, capital returns, and contribution margin accounting in the area of financial accounting, as well as turnover, sales, punctuality with delivery, delivery reliability, and so on, in logistics.

When the organizational structure of an enterprise is mapped onto the organizational units of the R/3 system, the business requirements must be brought into line with the technical system requirements. Defining system and cross-client organizational structures, information flow processes, and business processes, leads to a logical, overall concept for the R/3 system. Technical considerations such as performance, availability, maintenance, release upgrades, and so on, form, for the most part, the basis for applying this logical, overall concept to one or several R/3 systems.

The representation of an enterprise within a central R/3 system offers maximum integration with respect to the business processes and information flow processes required. The above-mentioned technical viewpoints, however, might not favor this kind of solution. When distributed R/3 systems are implemented, options to use distribution techniques, such as Application Link Enabling (ALE) or the Ledger Roll-Up technique of the application component FI-SL (Special Ledger), must be taken into consideration.

Figure 3. Several application servers acessing a single database

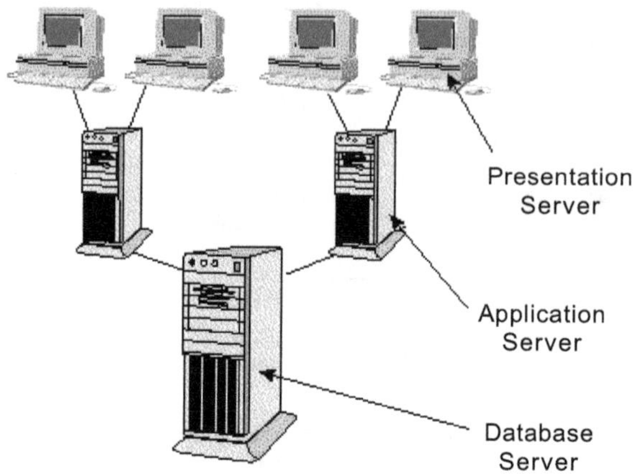

An R/3 system consists of three different levels (presentation, application, and database servers) in which each user may select different functions. The database server stores all data, as well as application programs, in a central database and executes all database functions (updates, and so on). Several application servers can access a single database server (Figure 3).

Dialog processing is executed in the application servers. The change requests are stored in the database asynchronously. To do this, the application server first reads a large data package from the database and executes the change requests within this data package. Afterwards, the data package is saved to the database as soon as the database server load allows.

The presentation functions, that is, the representation of the graphical user interface, are executed by front-end systems ("presentation servers," PCs, X terminals). With their help, data can be entered into and received from the system. An application server can serve a large number of front-end systems that are linked to end-users.

The hardware structure of an R/3 system, which consists of the database server, a number of applications front-end systems, and so on, plays only a minor role in mapping the organizational structure of an enterprise onto R/3 (Figure 4). Distribution scenarios that are based on ALE techniques are usually designed for an R/3 system to provide centralized functions in Financial Accounting and decentralized R/3 systems for the Logistics functions.

Figure 4. Hardware structure of an SAP R/3 system

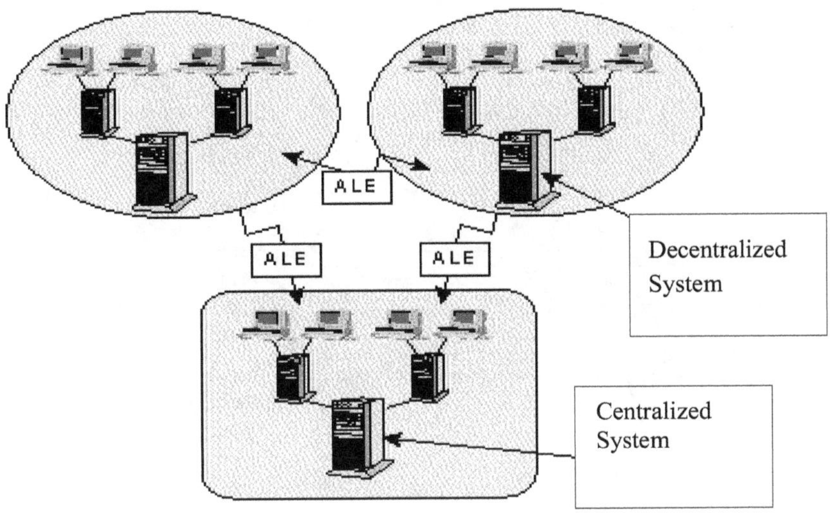

Logistics data that is relevant for Financial Accounting is transmitted as a single document or totals records from the decentralized Logistics systems to the central Financial Accounting system.

The range of possibilities for designing system infrastructures between a regional and a business unit-oriented system topology is extensive. As a rule, the system topology of an enterprise is a combination of these two approaches. For example, an enterprise in the chemical industry may set up independent R/3 systems for each of the following (Figure 5):

- each large foreign subsidiary
- each business unit
- each large production location

The business unit systems are the leading Logistics systems. Each strategic business unit contains at least one R/3 system. Larger business units with high demands with respect to performance and availability use several R/3 systems, often distributed over different regions. Materials Management, Procurement, Production Planning, and Sales are integrated into these systems, as they have already been executed for several foreign subsidiaries.

Figure 5. Set up of independent SAP R/3 Systems

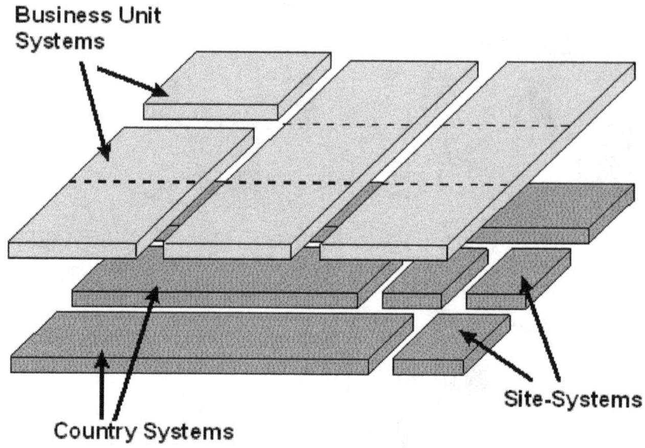

Regional services are integrated into the systems of the foreign subsidiaries. These services include, for example, time management, as well as the creation of reports for external accounting purposes for several business units. In the systems at the main location, services that pertain to the main location (for example, transport logistics, asset maintenance, and so on) are represented.

The link between business units and location systems on the one hand, and the foreign subsidiaries on the other takes place using the "ledger roll-up" technique. In this way, data that is required for setting up external reports is transported. The link between the business units themselves or between business units and main-location systems takes place via the ALE technique.

An enterprise could commence the implementation of R/3 with a central R/3 system for its main location. As soon as some experience with the operation of the R/3 system has been acquired, further locations could be changed over to R/3. After some years, a large business unit might be split off, not only in a legal sense, but it could also be made independent from a technical systems point of view. In such a case, the regional R/3 systems need to be able to cooperate with a business unit-oriented R/3 system.

Now that we understand how SAP R/3 approaches structuring an organization's information systems, we need to analyze how RB planned to approach the implementation of its new system.

Figure 6. Bosch's strategic position

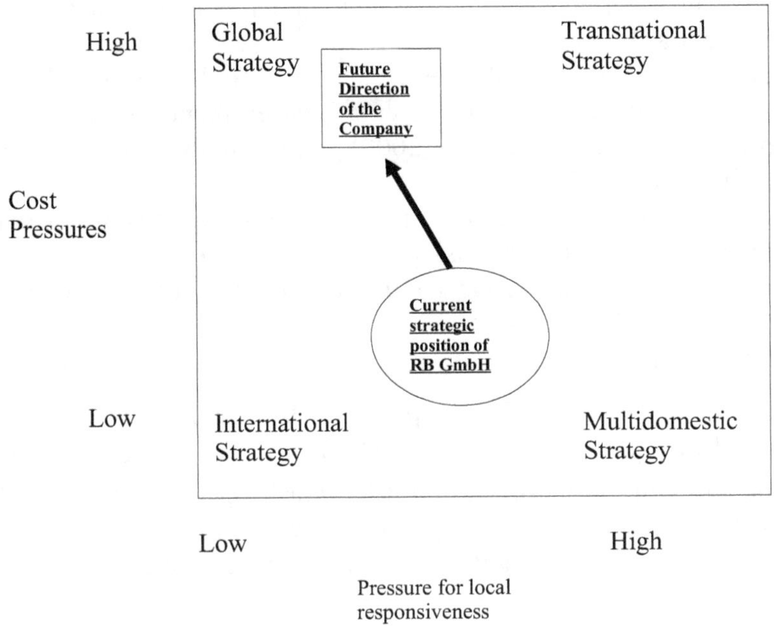

What Should Be the Approach to Implementation?: Domestic vs. Business-Unit Approach

The company is moving away from being a multidomestic company toward adopting a global strategy (Figure 6). The pressure for local responsiveness and cost pressures is relatively low at this time, leading to the classification of Robert Bosch GmbH as following a multidomestic strategy. In the future, however, the cost pressures are expected to increase and the pressure for local responsiveness will decrease as the number of suppliers decreases. These trends are guiding the company towards a global strategy. Figure 6 shows this shift in the company's strategic direction. Given the shift to a global strategy, the need for a standardized system such as SAP R/3 across all divisions of the organization has become critical.

The main advantage of this new orientation is that the company will be able to evolve all its divisions and operations together as an integrated entity. If this is accomplished, then many of the comments made in the case study (such as Kamran's vision of a fully integrated production and delivery system) become possible, and the new system will deliver a fully integrated customer service.

The disadvantage is that a standardized effort requires the full backing of the company's executives at both the corporate and domestic levels. This does not seem to be true for the company at present. Therefore, implementation problems might be severe. It might be worthwhile for the Board to sponsor an annual forum so that internal and external customers come to understand that the Board fully supports the standardization of SAP R/3 throughout the organization. The corporate management needs to define the strategic goals and ask QI to break them down to operational goals and create measurements to show process improvements. As Thomas Davenport (2000) points out, the implementation of an ERP system is primarily a business project and not an IT project. These multiple strands need to be tied together as a reporting system so that the performance of the different divisions can be easily monitored. The prerequisite of such a corporate-wide approach is that the similarities between countries and product divisions must be very high.

Implementation of Domestic ERP Approach for Each Country

The company is currently following a multidomestic strategy, as shown in Figure 6. Given this strategic direction, it might be worthwhile for the company to implement ERPs in each country independently, without insisting that they all work together. This strategy might be the best fit given the current corporate strategy.

The advantages of this strategy are that different countries will be able to use different systems depending on their investments and returns. For example, some of the divisions at RBUS might decide not to change their current systems, and this might be beneficial from the local business point of view. Even if QI were to provide economic incentives for the domestic units to standardize across the company, they may prefer this approach as it preserves the independence and autonomy of the countrywide operations.

The disadvantage is that it may not be feasible to close the books in five days, the Bosch board's stated goal, if SAP R/3 is not implemented fully across the divisions. The concept of top-down design, where the best processes are identified and implemented uniformly across the organization, is a good concept. A major issue is whether the end-users will be willing to buy into the best process concept, when for a long time they have been used to their own systems.

QI could influence the directions of the countrywide MIS operations by assuming the following roles:

- Be a leader in showcasing best processes and applications.
- Take into account country-specific MIS operations when identifying the best processes.

- Conduct international forums in as many different countries as possible to send a signal that QI's staff will be willing to travel to wherever they are needed for the implementation.
- Create methods and procedures to improve trust among the IT division employees.
- Provide training for employees so that they understand the benefits of implementing standard systems.
- Fire employees who do not agree to the approach suggested by QI.
- Showcase positive examples of how the implementation improves the efficiency of the company's operations, thereby improving the overall performance of the company.

It is also possible for this alternative to be implemented as an intermediate solution for Bosch.

Business-Unit-Based ERP Approach Within the Global Operation

Another approach is to have each business unit operate seamlessly with other business units that produce the same products around the world. A good implementation could provide Robert Bosch GmbH with numerous opportunities to enhance long-term profitability and growth. This approach would be the logical conclusion of treating the product divisions like enterprises within the enterprise.

The advantages of this approach are that it will be possible to close the books of each business unit within five days, thereby assuming full control of the operations. It would also make the employees working on the same product line feel like colleagues rather than competitors.

The sheer size of the Bosch group, which employs over 180,000 employees all over the world, makes it impossible to reach a consensus in favour of a common system. Likewise, the heterogeneity of the businesses in the group, each with their own different business processes, tends to indicate that it would make sense to have different solutions. The need for standardization could be reduced to some extent in the financial area. As Thomas Davenport (2000) reported, the business unit approach was followed by HP very successfully, even though the investment was higher than would be needed for a fully unified R/3 implementation.

The disadvantage would be that the business units within a country might not choose common systems, making it difficult to run them as integrated entities. Power is held by the executives controlling each business unit, rather than the countrywide executives. A change in organizational structure might be necessary to resolve this issue. To reduce this negative effect, the company should be organized globally in the sense that not all business units should be headed by German-based management. In this case, the company would take into account the needs of the local management in the different countries, particularly those who are not members of the executive body.

Having discussed the enterprise modelling methodology and the different options available for structuring the IT systems at RBUS, we will next analyze the different architecture options that were presented to the top management of RBUS so that an architecture option could be chosen for implementation.

Which Architecture Option is the Best for RBUS? Viewpoint of Mr. Lehmann

Gunter Lehmann, the coordinator of the SAP Base and Customer Competence Center (CCC) , is responsible for structuring and harmonizing the SAP system in QI. If he were asked to consider the impact of structuring the R/3 system in an organization and to explain the alternatives to the presented structure for RBUS, he might make the following recommendations for RBUSi:

The need to develop an IT architecture arises due to the following reasons:

- Long term investment requiring commitment
- Incorrect initial development cannot be overcome easily
    - 70% of business process reengineering projects fail (Lientz & Rea, 1999).
    - 70% of software engineering reengineering projects fail
- Properly implemented, IT architecture can integrate all facets of the corporation
    - Allows for increased efficiency
    - Can greatly reduce costs

ERP implementation projects fail for the following reasons:

- Lack of management commitment (no shared vision)
- Outside use of consultants
- Lack of methodology (do not know how)

- Lack of technical expertise (using old technologies, or no competencies)
- Lack of management follow-through (no long term commitment)
- Internal politics
- Focus on the work function (incremental) instead of the process (overall work flow)

Alternative One for the IT structure, presented in Figure 7 (also Appendix 1b, Chapter VII), provides the following features:

- Centralizes financial, human resource, and manufacturing systems
- Creates a common data structure at corporate level, forcing this structure throughout the organization
- Sets up separate instances of SAP for each division, allows for customization, within the data structure, of SAP system (different views)
- Consolidates data at RBUS headquarters before sending it to Germany
- Processes can be shared between divisions, but allows for the specialization that would be required without unnecessary compromises
- Requires interfacing separate instances through ALE or batch processes
- Interfaces between separate instances are likely to add considerably to the cost

Figure 7. Architecture 1

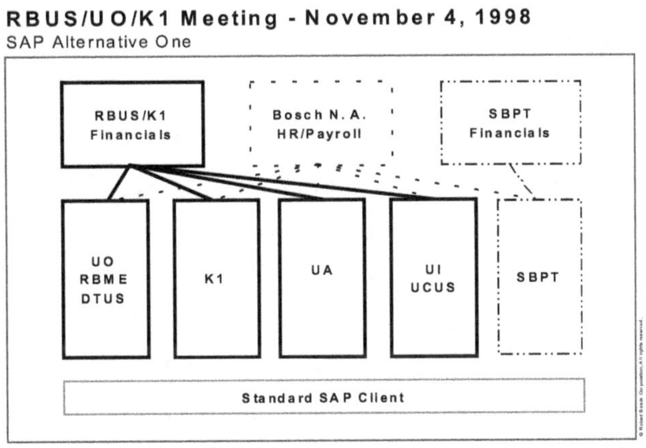

Figure 8. Architecture II

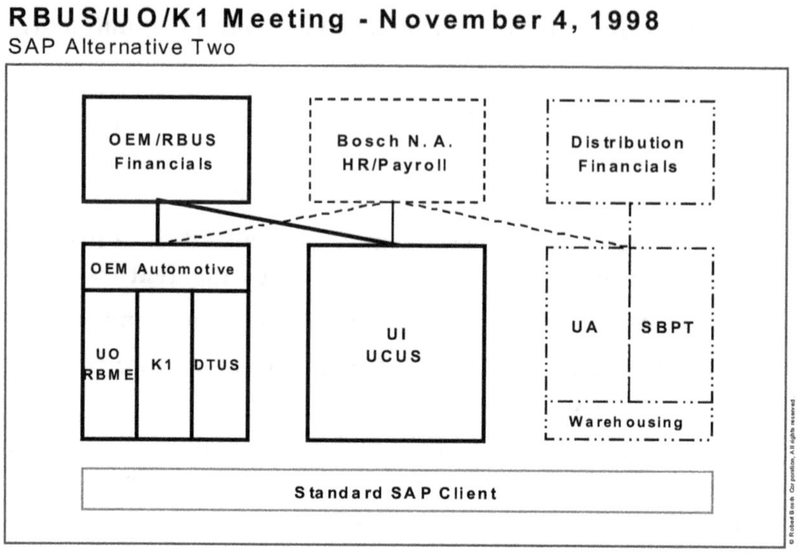

Alternative Two for the IT structure, presented in Figure 8 (also Appendix 1c, Chapter VII), provides the following features:

- Centralizes financial and human resource system, with decentralized manufacturing systems
- Creates a common data structure at corporate level, forcing this structure throughout the organization
- Combines instances of divisions with similar business processes
- Combine UO, KI, and DTUS as Original Equipment Manufacturers
- Combine UA and SBPT, since they share similar distribution systems
- Consolidates data at RBUS headquarters before sending it to Germany
- Processes can be shared between divisions, but allows for the specialization that would be required without unnecessary compromises
- Requires interfacing separate instances through ALE or batch processes
- Interfaces between separate instances were adding a lot of cost
- Batch processing allowed only from non-SAP systems
- Avoids significant cost and simplifies project

Figure 9. Architecture III

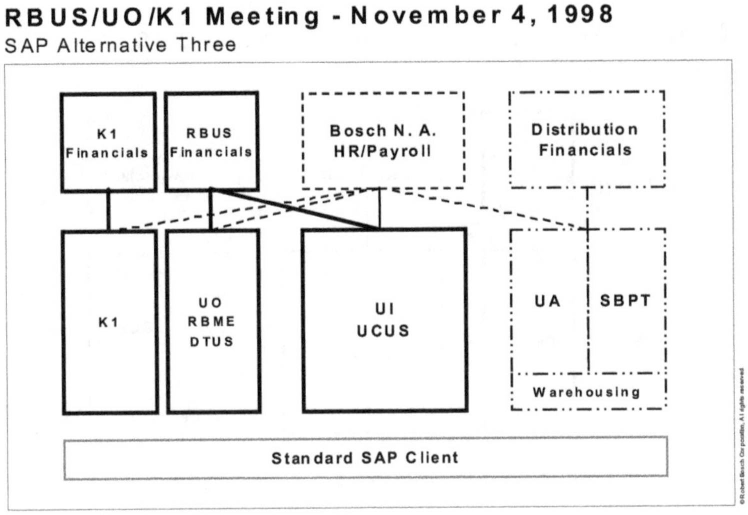

- Allows the cost of the project to be shared among divisions
- Might allow for the specialization of processes and enable more of the companies to make significant changes

Alternative Three for the IT structure, presented in Figure 9 (also Appendix 1d, Chapter VII), provides the following features:

- Centralizes the human resource system, with decentralized financial and manufacturing systems
- Creates common data structure at corporate level, forcing this structure throughout the organization
- Separates K1 from the rest of RBUS in the financials

  - They have already implemented an SAP R/3 system
  - Allows them to run their system as they see fit without having to integrate with the rest of corporate

Figure 10. Architecture IV

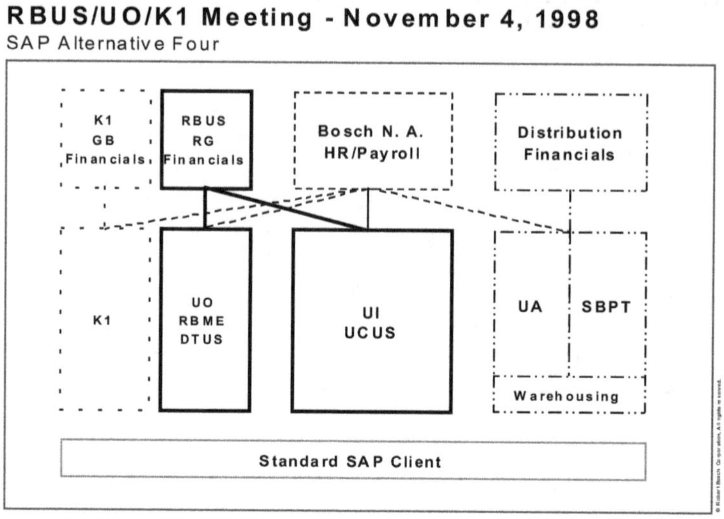

- Combines instances of divisions with similar business processes
- Consolidates data at RBUS headquarters before sending it to Germany

Alternative Four for the IT structure, presented in Figure 10 (also Appendix 1e, Chapter VII), provides the following features:

- Centralizes human resource system, with decentralized financial and manufacturing systems based upon the same SAP customization
- Creates a common data structure at corporate level, forcing this structure throughout organization
- Separates K1 from the rest of RBUS in the financials and connects them directly with the division GB worldwide
- Decisions for them would be made in Germany on a worldwide division level

The recommendation by Mr. Lehmann could be Alternative Two for the architecture (Figure 8: Centralized financial and human resource systems with decentralized manufacturing systems) due to the following reasons:

- SAP Alternative 2
    - Allows for long term extensibility and flexibility
    - Offers greater integration of business processes
    - Will take longer to implement due to cross-functionality between divisions
- Must implement both Financials and HR/Payroll modules
    - Interface issues/cost will arise

Which Architecture Option is the Best for RB? Viewpoint of Mrs. Bauer

Given that Margit Bauer is responsible for the software applications unit in QI and thus represents RB GmbH, her viewpoint will be biased toward what is the best for RB and not necessarily what is best for RBUS[ii]. Her analysis might be as follows:

QI has developed standards for:

- Organizational units
- Internal Bosch chart of accounts
- Automatic account finding in the R/3 system
- Balance sheet
- Structure of customer numbers
- Structure of vendor numbers
- Conditions of payment
- Payment terms
- Groups of goods

From the headquarters point of view, centralized systems yield a great deal of information that can be retrieved and analyzed very quickly. Therefore, IT architecture I (Figure 7: centralized financial, human resource, and manufacturing systems) is preferable. Using this architecture, it is possible to provide manufacturing operations with more flexibility and give them decentralized systems in the future (Figure 8).

Which Architecture Option is the Best for RB? Viewpoint of Mr. Chauncey

Don Chauncey represents RBUS and his preference is thus likely to be for the architecture that will be most suitable for enhancing the operations of RBUS[iii]. The recommendation by Don would be for IT architecture I (Figure 9) or for architecture IV (Figure 10). The reasons are as follows:

- By keeping the structure centralized, RBUS is able to keep their system centralized so they can better adapt to changes
- Upper management can make all final decisions
- Structure allows for specific knowledge of where information comes from
- Allows for more straightforward design, creating less chance for error
- Allows for more direct reporting

Don could also recommend architecture IV (Figure 10) since it requires the least amount of change to the organization.

The Need to Choose the Optimum Architecture for the Company

Given that the three teams represented by Mrs. Bauer, Mr. Lehmann, and Mr. Chauncey are likely to recommend different IT architectures, it is important to work together and come up with a IT architecture and implementation plan that could be presented to the Robert Bosch GmbH board.

It is important to note that architecture I (Figure 7) requires major changes to all the systems at RBUS. Architecture IV (Figure 10) requires the least amount of change, but the highest number of ALE links in order for the systems to communicate with each other. It is thus likely that QI will prefer architecture I (Figure 7) since it provides full control of the worldwide operations from Bosch's headquarters. The regional units might prefer architecture IV, since it provides maximum flexibility to them in finance and manufacturing operations. Only human resource is centralized and that would be helpful to many units since people frequently move between units. A systems architect might prefer architecture II (Figure 8).

The company executives have to work together to select an acceptable architecture option. It is possible to link the discussion with the literature on IT architecture discussed in earlier chapters. Once an architecture is chosen, then implementing the architecture requires a good understanding of the system topology and all the project management issues must be resolved.

What Should Be the System Topology, or the Physical Distribution of Applications to Different Systems?

The highest degree of technical integration is needed to ensure a "one instance solution." This means that all R/3 modules/applications of a company will be working with the same database. One restriction for this solution is the number of concurrent users. Within the three-tier architecture of the R/3-system it is no problem to multiply the number of application servers, but the bottleneck is likely to be the central database server. The performance depends on the system platform (e.g., UNIX-server, NT-server, OS/390-server), the number of users, and the quantities of transactions. A final decision depends on the amount of detailed data that must be stored. Therefore, it is generally not possible to make a final decision on this issue without a detailed analysis and, probably, performance tests.

From a business point of view, these problems with regard to the distribution of the logical system to different physical systems must be transparent. Thus, RBUS should define their enterprise model independent of the hardware structure in the first step. The physical distribution must be supported either by available ALE scenarios or Bosch-wide solutions. For example, from a corporate standpoint, it would make sense for the company's cash management to be centralized, allowing master data and control data standards to be managed centrally.

There is always a need for some financial functionality to be available in the logistic system of a plant; for example, a goods receiving transaction can be used to initiate a real-time change of the inventory account in the general ledger. Although it may be preferable for the HR application to be in a centralized repository to ensure a countrywide human resources policy, if a personnel cost planning module is used in a plant, then this system must be available for the on-site managers to access. In a one-instance solution, this is primarily a question of the level of authorization provided for each level of management, while in a distributed solution, it is also a question of passing necessary messages between the systems (or ALE scenarios). It may be necessary to investigate the requirement for a data-warehouse functionality. The corresponding SAP solution is the Business Information Warehouse System (BW-System), which is not part of the R/3 system but a separate system that communicates with the R/3 systems using an ALE interface.

How Should Multiple Systems Be Connected Using the Application Link Enabling (ALE) Feature of SAP?

The need for distributed business processes is demonstrated in the following excerpt from the case study:

The R/3 system at a company's headquarters, for example, may take care of central applications like accounting, human resources management, central purchasing, or sales planning. In local manufacturing plants, autonomous R/3 applications plan and control production and organize materials management, while other local R/3 systems handle order entry processing, shipping, and invoicing on-site in local sales offices.

It is possible to use the features recommended by SAP, as shown in Figure 11. Many enterprises are turning their attention to the challenge of loosely linking stand-alone application systems in order to take advantage of time- or event-driven exchanges of business processes. Some companies need to control communication between local systems in their sales offices and manufacturing plants, with central systems running from their headquarters. Other enterprises

Figure 11. Separation of Financials and Logistics (SAP AG, 1999)

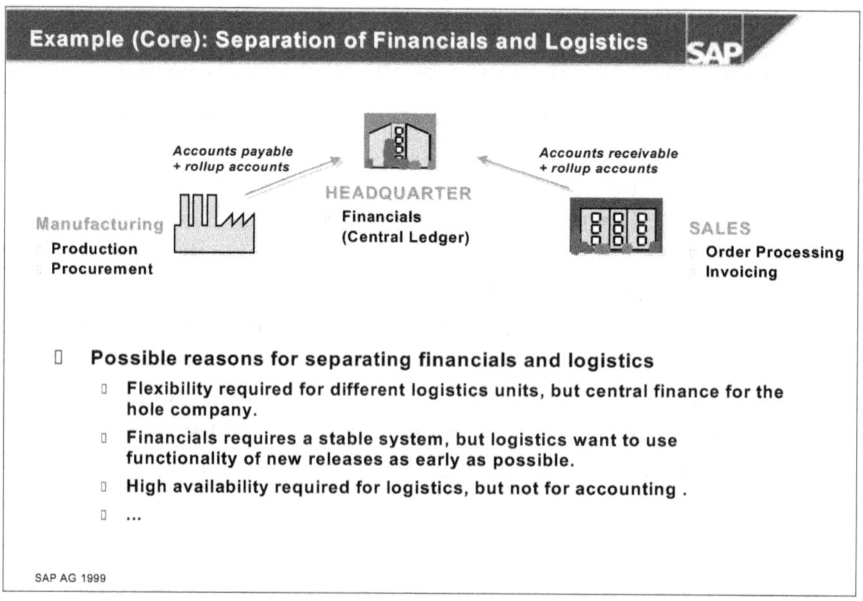

prefer to loosen linkages and improve the performance of tightly coupled systems. This means that while linked systems operate independently, they still remain loosely coupled with one another.

To meet these needs, SAP has developed Application Link Enabling (ALE) technology. ALE offers a new perspective in information management for distributed systems, linking distributed SAP applications with one another, as well as with applications from other vendors.

Application Link Enabling expands the strategic model to allow uniformity and controlled exchange of business data in distributed systems. Communication among loosely coupled applications is controlled by configurable distribution mechanisms. These mechanisms coordinate exchange of business information and update the master data among technically independent systems.

What are the Project Management Issues Involved in Implementing the SAP R/3 System for Robert Bosch Corporation?

An R/3 project should be considered primarily as a business-driven project. The senior management and the line management must be convinced that R/3, as an ERP system, is a tool that they can use to improve the market-oriented strength of the company. There must be a strong connection between the business strategy and the IT strategy. A prerequisite is to have strong backing from the company's top management for such a huge, important, and risky project. A survey conducted at the Harvard Business School revealed that fully two-thirds of responding executives believe that ERP has at least a moderate chance of hurting their business because of the potential for implementation difficulties (Austin & Nolan, 1998).

The sheer size and complexity of ERP initiatives are almost unprecedented in most companies. In addition, ERP initiatives involve both a major redefinition of traditional business processes and the company-wide adoption of new technologies (Austin & Nolan, 1998). This is not part of the daily routine for most managers. The high level of risk arises due to the lack of experience of both the users and the company's IT people with the SAP R/3 system. However, if the management has a clear overall understanding of the risks involved, then they will be able to look for solutions that will reduce the risk potential.

There are several approaches the company could take to implement a large group of applications such as financials, cost accounting and cost management, logistics, and human resource in different organizational units. At one extreme, the implementation could be done by introducing all the applications/systems in every unit simultaneously, a so-called "big-bang" implementation. At the other extreme, the new systems could be implemented "step-by-step;" for example,

Figure 12. Model for big-bang solution

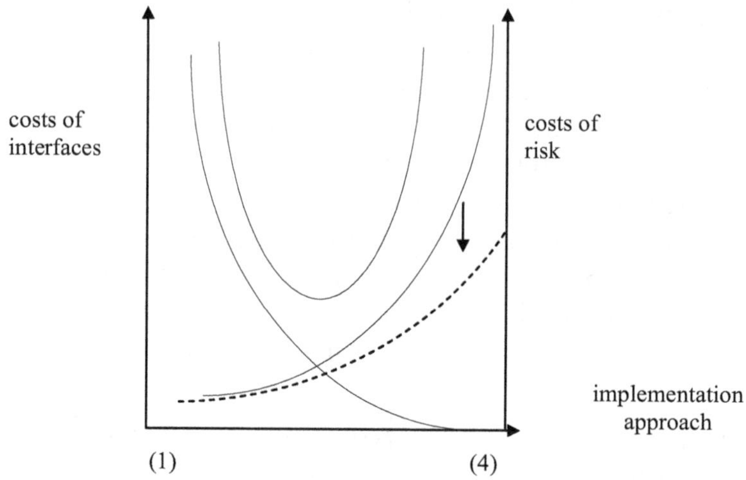

first for application A in unit 1, then application A in unit 2, and so forth, then application B in unit 1. This example implies a function-oriented implementation. Another approach could be process-oriented. The process-oriented approach follows the business process across the organizational units.

Irrespective of the approach used, the complexity and the risk of failure involved in implementing the SAP R/3 system is huge, but the potential benefits far outweigh the risk. The advantage of a big-bang solution is the avoidance of the need for temporal interfaces between the old and new systems. Figure 12 shows a graph depicting this interdependence. Along the x-axis there is a continuum, which progresses from a functional oriented step-by-step approach (1) to a process-oriented big bang (4). Table 3 defines the four principal categories, (1) to (4). In reality, of course, there is a whole spectrum of alternatives that fall between these four options.

Using this model, it is possible to determine the optimum strategy for a particular company, where the sum of both cost categories is minimized. If it is possible to reduce the costs of risk by using readily available knowledge and experience, the preferred solution tends toward the big-bang alternative (this situation is shown as a dotted line in the figure). In the RBUS case, such an implementation is possible due to the extensive experience the company has already accumulated as a result of the prior K1 division implementation in the U.S. and the in-house expertise available in QI.

Now that the architecture has been chosen, the implementation topology is understood, and a project management method selected, it is important to

Table 3. Strategy of implementation vs. orientation

	Step-by-Step	Big-Bang
Functional oriented	**(1)** Implementation of single functions at different times	**(3)** Implementation of all functions at once
Process-oriented	**(2)** Implementation of single processes at different times	**(4)** Implementation of all processes at once

understand the factors that will make it possible for the implementation to be a success.

How Can We Ensure the Success of the Implementation Process?

Austin and Nolan explain:

What seems crazy in a venture capital context, though, somehow seems normal in the information technology (IT) world of "Enterprise Resource Planning" (ERP). ERP initiatives are huge, often focusing on supply chain reengineering and integration, using products from companies like Baan, Oracle, Peoplesoft, and SAP. The promised benefits are too great to ignore: breakthrough reductions in working capital, huge bounties of information about customer wants and needs, and, most important, the ability to view and manage suppliers, strategic alliances, and customers as an integrated whole. But the costs and the risks involved are also huge. ERP implementation entails simultaneous and profound change in both business and technology. Each kind of change on its own is difficult enough. Attempted together, the job becomes almost unmanageable. Failure rates are high. (Austin & Nolan, 1998, pp. 1-2)

The traditional approach to managing an IT-project is based on a methodology that emphasizes the importance of planning. However, the first flawed assumption is that it is possible to prepare a comprehensive plan that will cover all the eventualities, including a list of the detailed requirements, and that the implementation is simply a matter of executing this plan. The reality is that a plan for such complex systems would necessarily be incomplete and imperfect because of the unanticipated problems that inevitably arise during the course of the implementation. The second flawed assumption is that a good planning process will include protection against late changes. The reality is that most businesses are characterized by rapidly changing competitive environments, so this may also cause problems if the plan does not retain sufficient flexibility to deal with new issues as they occur. The third flawed assumption is that it always makes sense to commit to a plan. Traditionally, this thinking should protect against the high costs of late changes, but this ignores the benefits of retaining some flexibility in order to be able to take advantage of any new opportunities that may arise (Austin & Nolan, 1998).

A new approach to managing an ERP initiative is to manage it like a new venture. A new venture has a probability greater than zero that the investment has to be written off completely. The management emphasis must be focused on the one hand on reducing this probability and on the other on realizing higher returns with a high probability. Austin and Nolan suggest that the following three factors are of primary importance when managing a large ERP implementation: staging, risk sharing, and people (Austin & Nolan, 1998).

- *Staging*: The paradox in implementing ERP systems is that because of the integrated nature of ERP products, it is necessary to swallow ERP more or less whole in order to reap the biggest profit. The cash flow profile of such an investment represents a bad business opportunity. To gain a commitment from senior management, it will probably be necessary to divide the investment into manageable portions. In most cases, the first stage should be to define an enterprise model and plan a step-by-step implementation process. This approach must take into account that each step must provide a visible and measurable business benefit without incurring an unconditional obligation to take the next step. The senior management must be given a real choice to either abandon the project or invest more money after each step. Staging an ERP initiative this way is by no means an easy task, but it is both useful and necessary. Of course, this staging incurs added costs due to the extensive interfaces needed, but it also reduces the risk of having to write off the investment and limits the potential damage, as well as providing a chance for gradual learning and the opportunity to use new information.
- *Risk Sharing*: If all the contributing parties lose when an ERP initiative fails, then all have more pressure and motivation to make it work. For example, Cisco included on the steering committee for its ERP implementation a vendor vice president and a senior partner from the systems integration firm. The payments to the systems integrators and vendors, as well as the bonuses for managers and team members, must be linked to the accomplishment of real objectives, for example a targeted inventory reduction. Also, the compensation of the senior managers and the affected line managers must be linked to the success of the initiative. Senior managers must clearly communicate to their own staff that this implementation will indeed take place, and will do so on schedule. Nobody should

have any doubt about their management's commitment to the project, or feel that they have their manager's tacit approval to drag their heels and delay the process.

- *People*: To make it clear how important the people involved are, there is an old business adage that goes: "I'd rather back an A-team with a B-idea than a B-team with an A-idea." Senior managers need to act as venture investors and line management who will enable the ERP implementation must be invested, perhaps through risk sharing arrangements. A strong project manager, excellent technical resources, and a good financial analyst must be supplied by the implementing company. It is imperative to obtain assurances from the vendor and the system integrator (consultant) that appropriate resources are available for the initiative and that they will be supplied in a timely manner.

The senior management, which in the case of RBUS is NAOC, must define its role as a venture capitalist. This is not simply a question of approving a well-prepared plan for the next two or three years, along with the necessary budget. In the complex and dynamic environment of an ERP implementation, a written plan may be obsolete by the time it is printed out. The challenge lies in creating well-defined stages, involving the management in a risk sharing position and having available the right people.

In summary, the steps to be taken to make the implementation experience a success are:

- Bring the CEO on board so that he/she feels that they own this project.
- Treat the implementation of SAP R/3 as a venture capital investment. Motivate IS staff, encouraging them to be innovative as they proceed through the implementation process.
- Report success stories.
- Plan the implementation of SAP R/3 carefully and be prepared to deal with the inevitable standardization and harmonization issues.

Chapter Summary

In this chapter, we analyzed the Robert Bosch case study from both an organizational and technical perspective. We discussed the alignment between the organization and the IT functions and how it could be improved further. The reasons the company decided to standardize on SAP R/3 architecture were also discussed. The enterprise modeling methodology of SAP R/3 was used to discuss alternative structures for RBUS, and then a set of four proposed architectures were analyzed. The possible strategies for implementing the architecture in the company were discussed and the big-bang approach was recommended as the best option for RBUS. The chapter concluded by discussing strategies that could be used to make the implementation experience a success.

References

Austin, R. D., & Nolan, R. L. (1998). Manage ERP initiatives as new ventures, not IT projects. *Working Paper 99-024 Harvard Business School.* Revised 12/1998.

Chandler, A. (1977). *The visible hand.* Cambridge, MA: Harvard University Press.

Davenport, T. H. (2000). *Mission critical: Realizing the promise of enterprise systems*. Boston: Harvard Business School Press.

Lientz, B. P., & Rea, K. P. (1999). *Business process improvement: Planning and implementation.* San Diego, CA: Harcourt Brace Professional.

SAP AG. (1999). *Online documentation, Rel. 4.0 b.* Retrieved on September 8, 2005, from http://www.sap.com

Endnotes

1 Our thanks to Sam Davidson, Joanne Li, and Van Norris for the analysis presented.

2 Our thanks to Barry Cumbie, Picandra Elzie, and Michael Hanes for the analysis presented.

3 Our thanks to Chris Dillard, Patrick Noll, and Satish Subramanian for the analysis presented.

Chapter IX

Robert Bosch SAP R/3 Implementation: Changes During 2000-2004

Learning Outcomes

- Understand changes to the organization during 2000-2004
- Discuss the status of SAP R/3 implementation for RBUS
- Understand the status of SAP R/3 implementation for RB GmbH
- Discuss future directions for the IT division at RB GmbH

Changes in Management During 2000-2004 in Robert Bosch as Related to the Case Study

During 2000-2004, many changes took place in Robert Bosch. In 2000, the CIO of Robert Bosch in Stuttgart, Dr. Eggensperger, resigned and was replaced by Mr. Gerd Friedrich. Mr. Friedrich holds a degree in mathematics and physics and spent the first six years of his career in the research and development department at Siemens AG. He then moved to Nixdorf, a fast-growing computer company in Germany, after which he joined the telecommunications division of Bosch, starting in R&D and ending as a board member before he was appointed as the head of the QI management team and as the CIO of Robert Bosch GmbH.

A few months later, Ms. Bauer left QI and became the CIO of the Bosch-Siemens Household Appliance Company. In 2003, Dr. Scholl, the chairman of the board of management, became the chairman of the supervisory board. Mr. Franz Fehrenbach became the sixth chairman of the board of management in the 117-year history of Robert Bosch. Dr. Dais, who among other areas is responsible for IT, became the deputy chairman of the board of management.

In 2003, Bosch's sales were Euro 36.5 billion, with a foreign share of 71%, and the company employed 229,439 people worldwide. The three business sectors, automotive technology, industrial technology and consumer goods and building technology, contributed 65%, 11.9%, and 23.1%, respectively, to the company's sales. A huge acquisition in the industrial technology sector reduced the dependence of the company on the vagaries of the automotive industry. The divisions within the automotive technology sector were renamed to become the Gasoline Systems (GS), Diesel Systems (DS), Chassis Systems (CS), Energy and Body Systems (EB), Car Multimedia (CM), and Automotive Aftermarket (AA) divisions. At a press conference in April 2004, Franz Fehrenbach, the chairman, stated that the company was "financially in good shape." He referred to the income before taxes, which rose from Euro 1.4 to Euro 1.8 billion, an increase in return on sales from 4% to 5% — according to Fehrenbach also due to the fact that "we made a major effort to improve our work processes." Across all business areas, Fehrenbach predicted the strongest prospects for Bosch in Asia and America, which is where he expected most of the economic growth in the next 10 years to occur. Asia alone was expected to generate more than half of the worldwide increase in automotive production by the year 2014. "This is a region of increased demand for us," said Fehrenbach. The growth rate in China for the company was expected to be about 25% on a Euro basis, similar to the level achieved for 2003.

In RB North America, the CEO Robert Oswald retired and his successor was Dr. Kurt W. Liedtke, who is also a member of the board of management. The Bosch Group in North America reported an increase in sales from $6.2 billion in 2002 to nearly $6.6 billion in 2003, with workforce figures remaining stable at more than 23,000 associates. Bosch automotive sales in North America increased from $4.2 billion in 2002 to $4.5 billion in 2003. Mike Bieganski was appointed as vice president of information technology for RB North America and was driving the process of implementing SAP R/3 in RB North America. Don Chauncey became the vice president for data security. Krish Kumar has left the company. Kamran Ashraf now works for Mike Bieganski as director of human resources and payroll.

Status of ERP Implementation at RBUS as of 2004

In August 1999, a proposal was presented to the NAOC to implement the SAP R/3 system. This proposal was put on hold by the decision-makers. However, the implementation of SAP R/3 in K1, which is now Chassis Systems, continued. In 2000 and 2001, a great deal of the standardization relating to data structures and business processes was completed. The automotive sector led the ERP implementation. Other sites were included based on the experience of implementing SAP R/3 in the chassis systems division. Finally, one R/3 instance for the automotive business was implemented with about 6,000 users across North America. This project cost $30 million. The company followed a phased approach, with the first step costing $13 million. During this stage, 11 sites were converted in 18 months, following an aggressive schedule to accommodate the elimination of data centers and legacy applications left over from business acquisitions. The second stage, costing $17 million, included conversion to the SAP ERP at six manufacturing locations, and an engineering/sales headquarters site, and also included implementation of the HR/payroll modules at all corporate and automotive sites (Figure 1). This figure shows that the SAP implementation was performed at the Automotive Chassis (AC), Automotive Powertrain systems — Diesel/Gasoline (AP), and Automotive Body (AB) divisions within the automotive business.

Figure 1. Priorities during 1990 to 2003

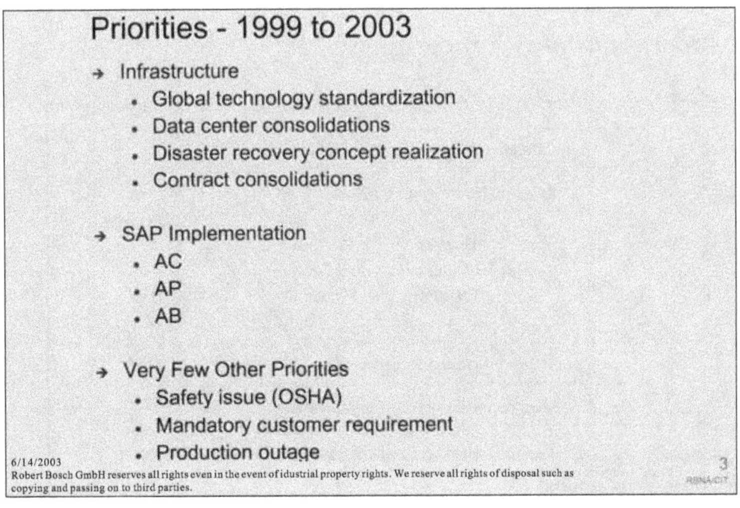

In addition to the implementation of SAP R/3, from a technical perspective they pulled the infrastructure together across the U.S. The data centers in both the automotive and non-automotive businesses were consolidated drastically, resulting in personnel reduction and cost savings as well as in improved quality of the processes.

Mike Bieganski shared his perspective:

The major success factor in accomplishing a project of this magnitude was to focus on the business process model instead of the technology (Figures 2 and 3). Although we knew from the beginning that SAP was not simply a technical implementation, we weren't prepared for how important it was to model the business. Our joint efforts with the business process experts revealed that there were four major value streams that could be harmonized across all divisions. This resulted in the decision to create one instance for all of automotive in North America, and would hopefully also enable future integration into a global model. For example, when reviewing the sales and distribution processes and the SD module in SAP, we realized that all locations were shipping products to common customers such as Daimler, GM, and Ford with common requirements. We harmonized the business processes, took best practices, and implemented. If one location had extra mandatory functions specific to it, then we included the extra functions as part of the overall model, making the functions available to other sites. So in North America the SAP implementation is a business, not technical, success story. We approached it from a business process perspective and used it to streamline our business and add value through integration. And

Figure 2. Macro business process

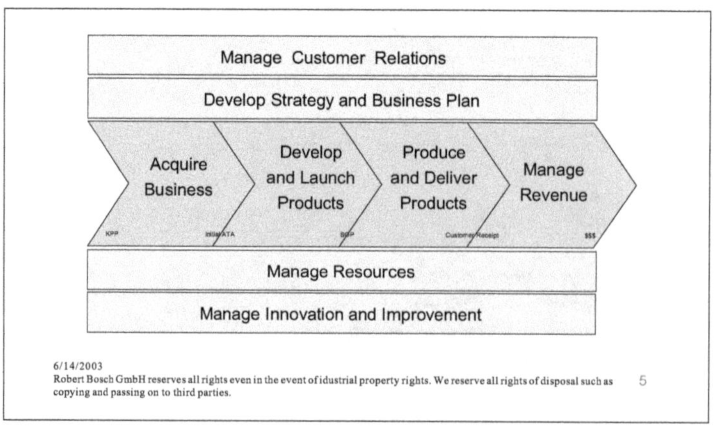

Figure 3. Business process and portfolio management

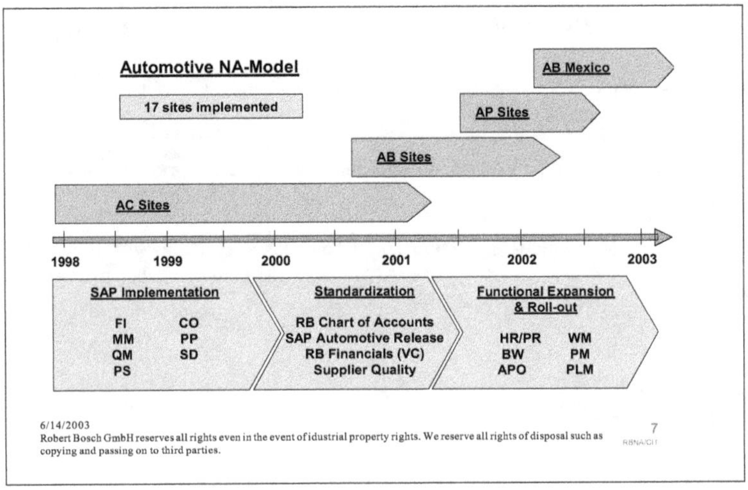

it really does help integrate the business when cross-functional resources are encouraged to communicate.

To implement such a major project, the right organization is critical. In order to manage the project, managers for the different areas, such as logistics, quality, and finance, were appointed. For integration purposes a program manager was named who focused on business processes and rather than having key users, business process leaders were appointed. By implementing these changes, the managers were able to match their business macro model to SAP modules. Mike Bieganski explained: "The steering committee concept was eventually dissolved in favour of the business unit liaisons within a portfolio management concept (Figure 4).

We like the principle of having business unit liaisons drive IT implementations from a business value-added perspective, instead of having steering committee overseeing projects. The IT team is now attached to all the major business process initiatives and has thus become technology enablers of the processes being driven by the liaisons (Figure 5)."

By mid 2003, about 13 legacy systems have been eliminated and this also made integration easier. The PRMS system and Computron no longer exist. HR/ payroll was also harmonized as much as practical, taking into account differences in union plant vs. non-union plants, local union rules, and differences in

Figure 4. Portfolio management

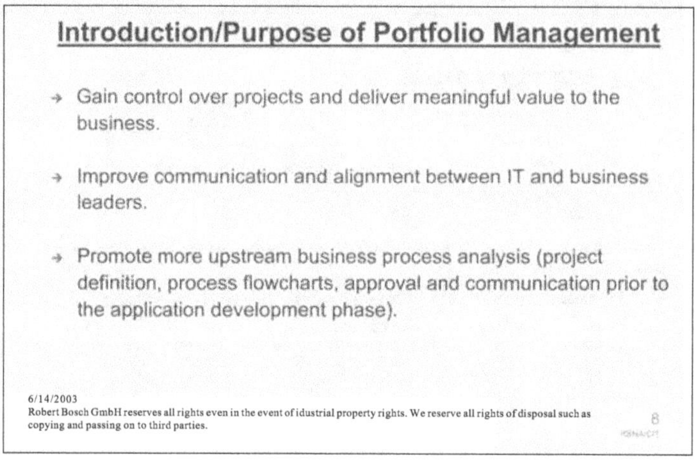

Figure 5. Management of IT projects

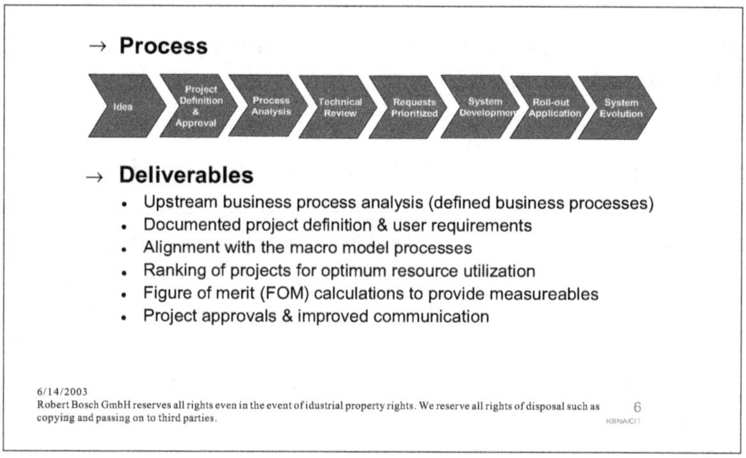

state taxes. However, huge benefits were realized due to the extensive harmonization. There is now one separate R/3 instance for all of North America for HR/payroll, integrated via ALE links to the necessary elements of the ERP system, such as cost centers, account numbers, and so on.

In the other business sectors, new SAP R/3 projects and initiatives are now in place. The business processes in the consumer goods sector were consolidated

Figure 6. R/3 systems architecture in Robert Bosch North America

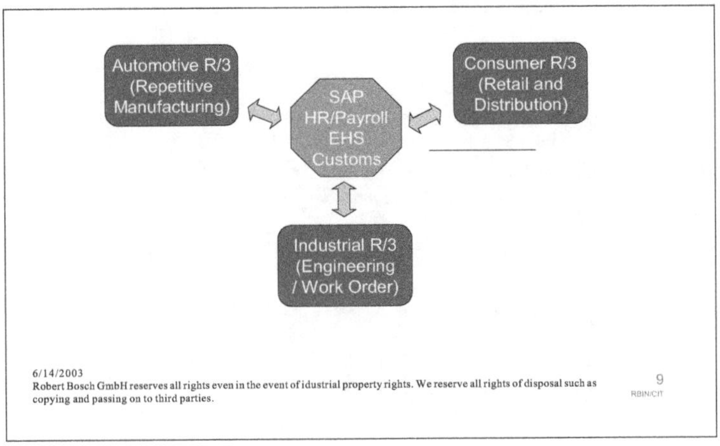

and integrated into the SAP R/3 instance. The industrial sector has its own SAP R/3 initiative. They are adopting the best approach of looking at the business strategy and business model and then enabling that through an integrated SAP system.

Looking at the discussions of possible system structures in 1999, which were really complex, and comparing those to the current systems architecture, the results show a relatively easier and clearer structure. Figure 6 shows that there are now basically four SAP R/3 systems interconnected by ALE scenarios.

Future Directions for IT in RBUS

The planning for future initiatives is now in progress. Mike Bieganski enthusiastically endorsed the process: "We know our business is global and we know we have to harmonize that business globally. We are trying to accomplish this through the QI-designed enabling layer to ensure that we can have regional views and also create a global model that reflects global processes. We believe it can work very well for us. And the portfolio management process will help us identify IT priorities and will also segment our initiatives into categories, allowing for appropriate resource allocation (Figure 7)."

Figure 7. Method of choosing future IT projects

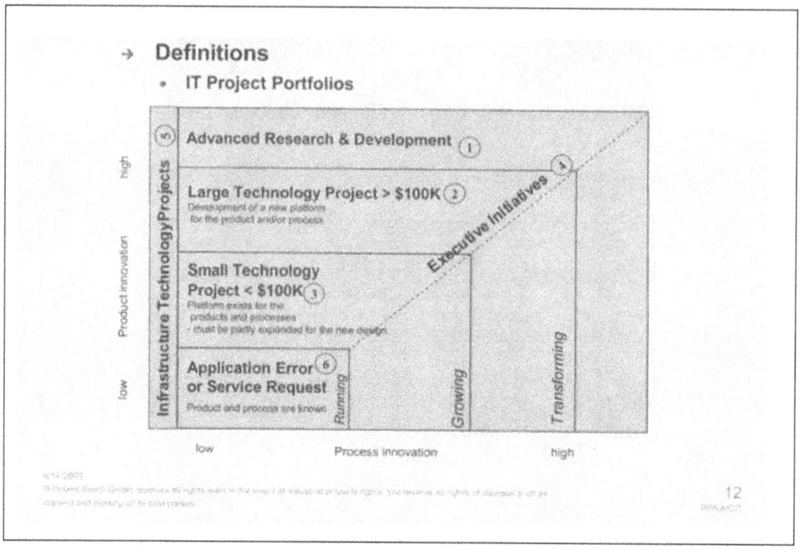

Status of ERP Implementation at QI as of 2004

During 2001, Mr. Harald Fontius became the vice president of program management of UBK, responsible for the R/3 project in the automotive sector for Europe and reporting directly to Gerd Friedrich, the CIO. Earlier, he had implemented an SAP R/3 solution in Brazil with five plants. Three major changes were initiated. These led to definition of a basic architecture for the corporation.

Change I: Focus Changed from Production (Plant) to Value Chain (Division):

A major issue arose as to whether a plant is the right unit for the ERP solution to be focused on. Harald Fontius explained the reasoning: "It was a major change from a business perspective. The plant is not any more the main focus for the management. The focus moved from 'how we produce' to 'how we support the whole lifecycle of products.' It is not important how many products a plant produces; it's more important how much value the plant adds to the product and that is controlled on product group level. The key performance measure earlier was the operating result of the plant. This has totally changed to a concept of value-based management." This implies that the company's key organizational unit has become the division, as it is responsible for the whole value-chain, starting with the product development, including manufacturing, and ending with sales. Convincing people that it is not necessary to have a plant system, but that there is a need to have a system based on the products offered by divisions, took about six months. It was decided that QI would offer a Bosch version of R/3 to each division. Such a version is an R/3 system where the Bosch processes are customized and could be used to roll out to a division. The division can then take the customized version and implement it at the plants. Meanwhile, QI is working on the next version by implementing the latest requirements. There is a company rule that each division does not have to implement every new version, but they do have to use at least every other version. This ensures that versions used in different divisions do not differ too drastically, while at the same time the divisions are not overloaded with too many version changes. This was a major change compared to the RB/3W initiative embarked upon during 1999, which focused on standardizing processes across plants (Chapter VII).

Change II: Increasing Importance of Process Harmonization

The next major change was the concept of process harmonization. This meant that QI had to identify and realize business processes, that were standardized and harmonized by the business processes owners across the divisions, such as chassis systems, diesel systems, and so on. Harald Fontius explained the concept of harmonizing a business process as follows:

If you talk to people in different divisions, they will tell you the purchase process is absolutely different in their divisions; one is dealing with 'purchase orders' whereas the other is dealing with 'schedules.' So at first glance it looks fairly different, but if you compare it to the functionality provided by SAP R/3 you could say these are two different options in the purchasing process that will be offered in the Bosch version of the R/3 system for this division. Often there are differences in the names used, but if you look into the details there are a lot of similarities. Sometimes the challenge is to bring the people together and get them to agree on a standard.

To handle the process of harmonization, employees from each of the divisions were recruited. For instance, for the purchasing process each division was represented in the purchase process harmonization team by its purchase process owner. To cover the changing business requests, a change request process was initiated, so that the harmonization team could discuss and decide how a request fits into the harmonization strategy and how the requests should be implemented according the version policy. Beside the harmonization of the systems that are used in the divisions, this allows people to speak the same language. For example, a purchase clerk from the Reutlingen plant, producing semiconductors and engine control units, can talk to his colleague in the Hildesheim plant, where car radios are manufactured, and get help in using the SAP R/3 system and fixing a problem. They understand each other even though the products are different, and are able to help each other. From a company point of view, moving people from one division to the other has also become much easier.

The change in the SAP R/3 implementation had a major effect on the organizational culture and management. Currently, the division is fully responsible for the whole life cycle of their product lines. So the plant today plays the role of a 'hotel' for the manufacturing facilities, which are assigned to different product lines. As far as possible, machines are assigned to product lines, so each plant is virtually separated into product lines. The plant manager is no longer responsible for manufacturing; this is the responsibility of the product manager within the division. This was a major change in the culture of Bosch, where the plant management was historically very powerful. Harald Fontius noted that those divisions who appointed former plant managers as product managers had fewer problems compared to those who appointed new product managers.

A question might arise: Why should IT managers worry about reporting by plant or division? The SAP R/3 system plays a major role in this organizational change, since in the new structure the plant manager is no longer receiving the reports he was accustomed to getting for the last twenty years. No plant reports exist any more, so the question comes up, 'how can I manage my plant?' QI had to deal with the plant organization's problem to fully understand that the different parts of the plant will be managed by the different product managers of the divisions, as the plant now consists of a number of product lines. Those activities and machines that cannot be assigned to a particular product line are integrated in one service area that provides services to the product lines and the product lines are charged for these services. Looking at this, the division now has the information available to decide where they will produce each product. They may decide based on whether the value added to the product is high or low, or look at other factors. For example, one option may result in a lower labor cost but a high shipping cost, and the shipping does not contribute any value to the product. A common rail pump could be produced in Germany, Turkey, Brazil, Italy, India, or Japan, as all these plants work closely together in a big manufacturing network. The division may decide to supply the U.S. market from the plant in Brazil, with only peak demands being covered by the Bamberg plant in Germany. Decisions such as these show what makes an ERP project complex and difficult. From the top-down perspective, the directions and strategies might be clear, but often it only becomes real on the working level when the new ERP system is implemented and a lot of organizational resistance issues are addressed.

Change III: One Development System for the Rollout:

The third major change was that QI decided to adopt only one development system for the entire automotive business and take this as the basis for the rollout to the divisions. Before Bosch went through this harmonization process, the implemented systems looked similar but were actually quite different from each other, leading to a set of incompatible development systems. For nearly every R/3 production system QI established, a separate development system was in place, making standardization impossible and maintenance expensive. This led to divergence.

With regard to the structuring issue within the R/3 system, Bosch decided that each division will be represented by a separate R/3 company code and have a separate R/3 instance; the diesel systems is a company code, gasoline systems is another company code, and so forth. Each plant in every European country has its own company code. For a big plant like Bamberg, products for several different divisions may be produced. Bamberg will thus have an R/3 plant code in the diesel R/3 system and another plant code in the R/3 instance of the gasoline division. From the R/3 system viewpoint, it was decided to handle the common services of the plant facility via the ERP solution. The typical solution is that one of the divisions will be charged for this service unit in its R/3 system. Via an internal customer-supplier relationship, the cost has to be charged to the other division. To charge this from one R/3 instance to the other automatically creates some complexity and systems experts are currently working to come up with a solution. Robert Bosch uses project management systems in order to implement the IT systems and roll out projects. Figure 8 provides an example of a roll-out plan used at Robert Bosch for Europe. This figure illustrates the scale and complexity of the project management efforts involved in a project of this magnitude.

Defining a Basic Architecture

Bosch realized during 2001-2002 that standardizing master data is essential so as to enable a high degree of integration. Thus, a finished good in one division has

Figure 8. Masterplan roll-out for UBK

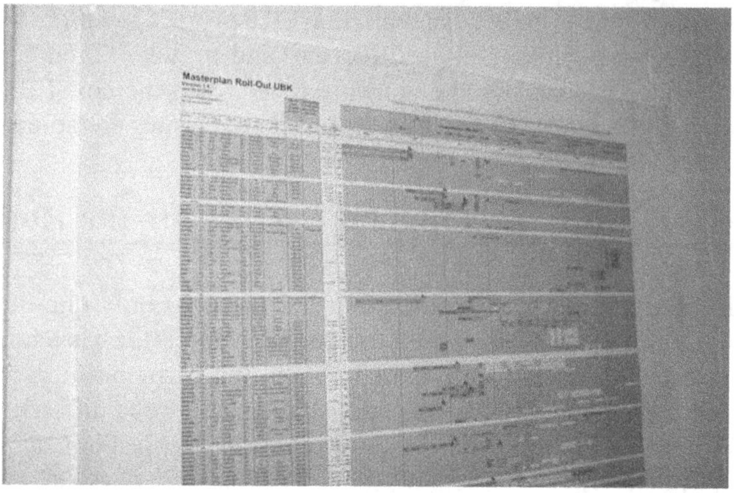

a single material number in the R/3 system. On the purchased goods side, a complete standardization cannot be achieved on a global level with all divisions. Bosch is using the classification system of R/3 to minimize as far as possible having different numbers for the same raw material. This works well for materials where contracts can be negotiated centrally. To support this scheme for important materials, for instance copper or certain semiconductors, global purchasing teams were created which collect all the demand figures and also consolidate the master data. To support and improve this operation, one of the new projects is to implement a PDM/PLM (product data management/product lifecycle management) system. This example makes it clear that architecture of the system is an important consideration in ERP implementation; otherwise, it will not be possible to manage using this system.

So far, we have discussed the European ERP project, where Bosch is on track to get a high degree of integration. If we look at Bosch worldwide, including its operations in regions such as Australia, India, Brazil, and the U.S., there are several different SAP R/3 solutions, mostly harmonized within each region. A good example for this is the U.S., where we have described the system that Mike Bieganski has implemented. However, to manage a division worldwide, one worldwide system on a divisional base is necessary. Bosch realized that doing this as one integrated system does not work for various reasons. Plants in different countries work differently. The conceptual idea is to add layers of integration. One layer is the Division Umbrella Worldwide, with a system for

Figure 9. Outline of building blocks of the systems architecture

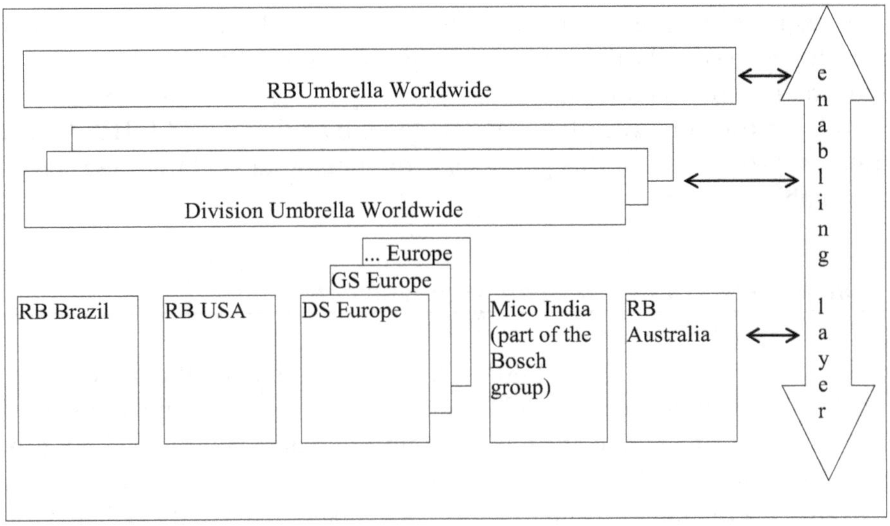

each division, and another is the RB Umbrella Worldwide layer. To combine these layers with the regional systems, QI uses an Enabling Layer (Figure 9).

Global Issues

Looking at this structure, the current and future challenge is to link the regional systems and to define which items (e.g., master data, codes, etc.) must be defined on which level and distributed to the subsequent systems. Another point Harald Fontius mentioned was that the divisions need to have an informational base to manage their divisions worldwide, but up to now the divisions in the different regional systems were an integral part of the regional companies. For example, the president of the Bosch subsidiary in an Asian country considers that he is responsible his company; therefore, the legal unit has to have one company code in the R/3 system and the division is represented as a profit center. The division manager, however, insists that he is responsible for the division as it is standardized in the European system, so he needs a separate company code for his division in order to get the necessary information. Basically, it is the same issue as it was in Europe regarding the relationship between divisions and plants. The top management now understands this issue better due to their experience implementing the IT architecture elsewhere, and is working to resolve the issue.

From a business standpoint, the European IT project now covers 70% of the business, with 30% covered by mostly harmonized systems on a regional level. This is especially true for Brazil and the U.S., where the implementation approach was basically the same but customizing of the systems is different. Based on the architecture shown in Figure 9, it is now possible to obtain consolidated financial results on an RB level by using an SAP BW (Business Warehouse) solution. For this purpose, a lot of mapping and linking is provided by the enabling layer.

Alignment Between QI and Top Management

The changes initiated in 2000-2001 were successful since the direction of the board of management relating to IT had already changed tremendously. When Gerd Friedrich joined QI as the CIO, he improved the discussion about process standardization and ERP implementation strategies with the board of management of Robert Bosch. The main strategy was discussed and firmed up with two members of the board, namely the CFO and Dr. Dais, who is responsible for IT. He had to convince them that it was necessary to separate the harmonization process from implementation and rollout, and to implement a separate project organization. In order to reduce the complexity, the project focused on the automotive business, which comprises about two-thirds of the Bosch business. The board wanted to start there because the advantages expected are substantial. To convince the board, it was essential to have sufficient backing to take it to the next level. This top-down approach was essential in making the project a success.

Bosch defined three different organizational entities to be in charge of: (1) defining, standardizing, and harmonizing the business processes and data structures; (2) implementing the processes in the R/3 system; and (3) rolling out the product in the different sites. The different divisions were invited to send representatives to work with the first group in order to define the business processes. If they chose not to participate, they were forced to accept the processes as they were defined. Based on the processes defined by the first group, the IT people in the second group implemented the processes in the R/3 System to produce a version that is ready to rollout. The third group is responsible for the education and training of all the people impacted, setting up the production R/3 instance for the division, and so on. Less than 20% of the total costs for the ERP systems are spent by the IT implementation organization; the funds are used by the other two organizations to create change in the company, change business processes, educate employees, and so forth. These three groups are coordinated by a board that reports directly to two members of the board of management.

Gerd Friedrich, the CIO, is a member of this board that coordinates the three organizations. He is, in addition, responsible for the IT implementation. Therefore, the CIO is not only responsible for IT, he is also (via the board) responsible for the whole process (defining, implementing, and roll-out). If anything goes wrong, the CIO is immediately aware of the problem. Gerd Friedrich commented that he knows of no other big company that operates this way. He believes that this helps the business units cooperate with each other to achieve the goals of the company.

Return on Investment for the SAP R/3 Implementation Project

During 2002 to 2008, Robert Bosch GmbH plans to spend Euro 750 million on implementing and rolling out standard processes in Europe. They also plan to spend an additional 15% on implementing the system worldwide. Adding these figures, the remodelled system is expected to cost approximately 860 million Euros or one billion U.S. dollars. They are currently struggling with calculating a return on investment, as it is not obvious how the divisions should do this in a transparent way. Gerd Friderich explained the problem:

If you believe in Microsoft's arguments that by implementing a new operating system you save 15 minutes per day due to the new functionality to store and find documents, and multiply it by 200 workdays and 10,000 employees, you get 500,000 hours a year, which are savings of about 300 people. Of course, these are not real savings in costs until the company terminates the employment of these employees. We can use such calculations to show an ROI for the SAP R/3 project, but we prefer not to do so. Our divisions are trying to find or build a valid model based on processes on how to calculate the ROI in a reliable way so other divisions can apply it. During 2001, Bosch was in the position that they were at the end of the life cycle of their major IT systems. They had to invest in IT systems in any case so as not to hurt the business, and that was the reason that the board of management accepted this investment without a sound ROI calculation. Normally Bosch has very strict rules for getting investment projects approved.

Gerd Friedrich went on to discuss the larger picture:

The big issue in this context is not the implementation. Of course, there are a lot of problems to be solved, but this is at another level. Change management is the most important issue; you have to change the mindset of the people. The board does not accept excuses not to use the standardized and harmonized processes. If a division is not willing to accept the standardized processes, we will bring it up to the board and then the division has to show figures that make it clear that their business will be damaged if they use the standardized processes. It is not only a question of whether they like the standardized systems or not — they really have to show the negative impact on the business by using a standardized process. The board has come up with such a process to resolve disagreements, but I have to admit that up to now no division has used this escalation process.

He referred to company-wide discussion about values, targets, and respect for its employees. In this context, there exists a common understanding that everyone will follow decisions that are made corporate-wide and will implement the decisions. Having this common understanding makes it easier to implement the changes and systems. During October 2003, QI had a meeting with the board to discuss the IT strategy of standardization and harmonization. The result was written in the minutes of this meeting and all the members of the board of management signed off on it. As a result, QI effectively has the contract that is essential for them to proceed with the SAP R/3 implementation.

Future Directions for QI

Gerd Friedrich gave his view of future developments:

Today QI is responsible for IT for Europe, Latin America, and Northern America, and the next step will be to include Asia in the QI organization. A managerial issue for the future is whether a region will have more to say than a division. There are discussions on the level of the board of management. This decision will influence the structuring of the systems heavily. The harmonization worldwide may not need much monetary effort. For example, with Robert Bosch North America they already have a harmonized system for the U.S., so it is easier to harmonize this with the European system. The same is true with Latin America. In Asia the business is starting to grow; Bosch will not have the issue of the migration of existing systems in that region.

Another initiative is to extend the ERP solution to other business sectors. QI has to construct building blocks as options to suit the different businesses.

When the European project is completed in 2008, the IT solutions will be enhanced further. Bosch is working on the next generation of ERP systems, and also plans to extend the functional scope, as the current roll-out does not cover all the features needed.

Bosch did not spend enough effort thinking about life cycles of IT systems in the past. Gerd Friedrich reflected on this: "If Bosch is developing a product, let me say a new injection technology, Bosch plans the life cycle of the product very carefully. This has not happened consistently with IT applications. When the end of the lifetime of a system arose, the discussion about future systems arose. So, if you don't have a lifecycle management for IT applications you normally start too late and you have too much to do in a short time span. We have implemented a strategic IT planning process which covers an eight-year period, so we have a roadmap for eight years." Using such a strategic planning process, Bosch is now much better prepared for aligning its IT and business strategies.

Chapter Summary

In this chapter, we describe the implementation of SAP R/3 at RB GmbH and RBUS during 2000 to 2004.

A centralized architecture was used at RBUS and one R/3 instance for the automotive business was implemented across North America. They harmonized the business processes, took best practices, and implemented them. A portfolio management concept was chosen where business unit liaisons drove the IT implementation. Four SAP R/3 systems, namely automotive, HR, industrial, and consumer segments were implemented and were interconnected by ALE scenarios.

Major changes in SAP R/3 implementation took place at RB GmbH also. QI decided to adopt only one development system for the entire automotive business. A PDM/ PLM system was implemented so as to standardize data elements. The changes have resulted in the focus of the company shifting from "how we produce" to "how we support the whole lifecycle of products." During 2002 to 2008, RB GmbH plans to spend Euro 750 million on implementing and rolling out standard processes in Europe.

Chapter X

New Directions in SAP R/3: Service Oriented Architecture (SOA) and Netweaver

Valentin Nicolescu, Technische Universität München, Germany

Holger Wittges, Technische Universität München, Germany

Helmut Krcmar, Technische Universität München, Germany

Learning Outcomes

- Be able to provide an analysis of the case study regarding process design changes and integration with other software systems in the future
- Understand how to simplify the design and implementation of cross-component processes
- Be able to utilize SAP NetWeaver as an IT toolset for realizing service-oriented architecture
- Be able to identify basic ideas for enhancing the given architecture at Robert Bosch Group based on the NetWeaver components

New Directions in SAP R/3

Review of the Current SAP R/3 Architecture

The case studies discussed in earlier chapters showed how the Robert Bosch Group (RB) implemented a structured ERP platform by using SAP and standardized most of its business processes, thus facilitating smooth information flows throughout the corporation. But what will the future bring and what issues still remain to be solved? Instead of simply providing predictions on the future of the ERP systems at RB, we will first analyze the current architecture.

The original problem at RB in the mid-1990s was a non-integrated system landscape based on outdated business systems that were not able to support future business requirements. As they considered how to deal with this problem, RB realized that the situation represented a great opportunity to completely redesign the landscape, which would better satisfy their long-term corporate requirements. After examining the situation from all angles, issues arose which involved more than just IT and organizational matters. The decision to implement the SAP modules using a single instance concept was accompanied by questions related to the existing structure of the organization and process flow. A complete reengineering of the business processes and organization was necessary in some divisions as a result of changing important IT systems.

When we summarize the results achieved by the implementation of the ERP system at RB, we see on the one hand changes in technical domains and on the other modifications in concept. The main technical achievement was the abolition of the legacy systems that hampered the development of new concepts and integration. By replacing the old systems with a single ERP system, the user interface also changed. This was one of the most noticeable changes for end-users because instead of using many different programs, only one interface was in place. Figure 1 shows that by segmenting the SAP systems into its most important parts and by adding the SAP user interface (or SAPGui), this "front-end integration" was achieved. In addition to this step, another important integration was realized, namely data integration. Important changes in the way business data was handled were synchronized with other ERP instances to avoid the problems caused by data redundancy. Later in this chapter we will explain why this kind of presentation was chosen for the R/3 architecture at RB.

Data integration is closely linked to data consolidation. It is important, therefore, to provide clear guidelines on how master data is to be created and how normalization procedures should be applied. At RB, these conceptual modifications took place not only in terms of process redesign, but also in the company organization structure. The standardization of both the processes and the organization were necessary steps toward global unification.

Figure 1. Technical view of the integrated RB SAP landscape

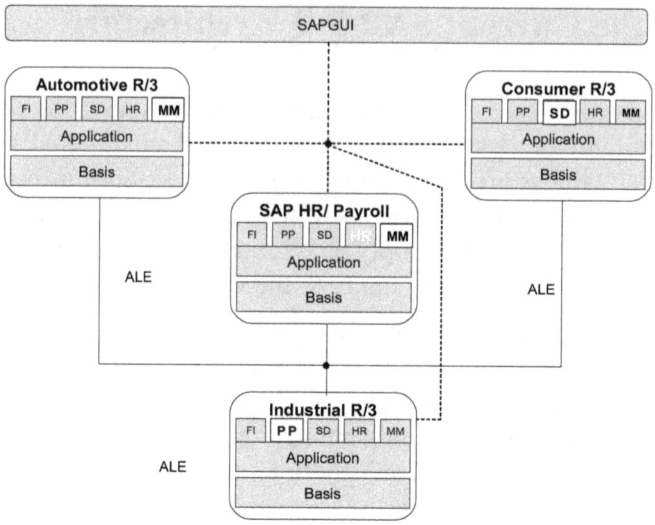

New issues may arise based on the ERP structure. For example, the question of enhancing the ERP landscape through the utilization of other SAP products, such as customer relationship management (SAP CRM), could emerge in the future. Another issue is the measures that need to be taken to integrate a non-SAP product into this homogenous architecture.

Chapter Structure

Some interesting issues are likely to arise after the implementation of SAP R/3 at RB in the future. This is why we will discuss a new approach called service-oriented architecture (SOA), which challenges the traditional approaches to dealing with the interaction between IT infrastructure and business process design. After presenting the general idea, we will take a closer look at SAP's implementation of this concept in the enterprise service architecture (ESA). As this concept needs a firm base to build on, we will include an explanation of the toolset that SAP proposes for realizing ESA: SAP NetWeaver (Karch & Heilig, 2004). This toolset contains software components that are designed to meet specific demands. We will then discuss how the application of these components could change the processes in the RB SAP R/3 infrastructure. Finally, we will pose some questions that will help you analyze how the use of NetWeaver could impact possible future implementation at RB and what the application of NetWeaver may mean for enterprises in general.

Future Issues

Please note that the issues that we present in this section are fictitious. Although not covered specifically in the case study, a discussion of these issues provides us with an opportunity to speculate on what the outcome would be if they did arise, and to propose possible solutions.

The first step taken by RB toward integrating their ERP environment was to implement a unified system. After the decision was made to implement SAP, RB wanted to protect their investment and took actions to enhance this architecture instead of implementing new systems from different vendors. As SAP does not provide all the software programs needed by a large corporation, the issue of integrating third-party or homemade products gains in importance. The integration of different applications is not only a problem for non-SAP products; the range of products offered by SAP has grown tremendously in the past few years, so the interactions between SAP applications have also become more complex. As RB plans ahead in order to maintain its competitive edge in a changing market, the ERP platform implemented must permit expedient changes in business strategies to be performed using an integrated infrastructure (Figure 2).

Some steps in the implementation process of SAP R/3 at RB (for example, the unified roll-out of processes) have already taken place. Nevertheless, some issues in different domains still need to be reviewed, and it is helpful to distinguish between the domains of process and data flow, data structure, and front-end integration. It is obvious that process integration is the only nontechnical issue in this list. Although Carr (2003) states that IT doesn't matter for business, we argue that it plays an important role in enabling companies to stay competitive in today's constantly changing business environment.

Figure 2. Enabling business strategies to align an ERP platform as an information system (Krcmar, 2005)

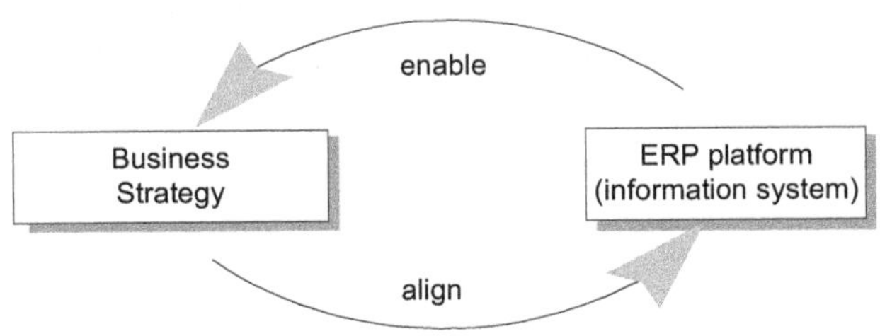

Process Integration

Process integration can be divided into two aspects: design and implementation. First, we will examine the design aspect for RB. We have seen that designing and distributing standardized processes is a complex task, even when a homogenous platform such as SAP R/3 is involved. Imagine now that RB decides to modify a crucial process. After modeling the new process flow, the customization of the affected modules in R/3 will need to be changed and distributed to all the plants. This entails a considerable effort, since the business part of the process needs to be changed and the technical settings also need to be adjusted to the new environment. As a result, even minor changes require the attention of both business and technical experts. Another scenario could be the design of a cross component process. We know that RB uses SAP R/3 as well as similar products, such as SAP APO (Advanced Planner and Optimizer) and SAP PLM (Product Life Cycle Management, which was discussed in Chapters IV, VI, and VII). What is the best way to implement a process that uses functions from both R/3 and APO? Bear in mind that communication with other SAP systems is relatively simple compared to the challenge of including all suppliers in a business process. Questions to think about are: Is it necessary to design the process anew for each supplier? Does the company need to make technical changes in all of the ERP instances for each new process?

Data Flow Integration

In the RB landscape, ALE connections are widely used as they are the most suitable solution for pure SAP environments. Furthermore, a central system is involved (human resources/payroll) where all the connections run together. This central system performs well as long as there are not too many ALE connections in the star-scheme. A central installation has the benefit of recording all data exchanges in the SAP architecture, but how will this landscape behave if more systems are connected to it? Performance problems might be encountered in the central system, or there may be many direct connections that can no longer be controlled from a centralized point. Additionally, proprietary interfaces for connecting third-party or homemade systems are also needed. If each system were to be allowed to exchange data with other systems, the number of interfaces would grow exponentially. Therefore, what solution reduces the number and cost of interfaces yet provides good performance and permits central process overview?

Data Structure Integration

RB has set up guidelines for the creation of new master data (e.g., customer data). This organizational measure unifies the naming of new customers and has thus initiated harmonization of data. The use of ALE connections between SAP systems makes it possible to synchronize customer data to a certain extent. However, this does not avoid redundancy of data, which can lead to different data sets. For example, if RB decides to use SAP's Customer Relationship Management system (SAP CRM), customer data will be stored in both the CRM database and the R/3 database. This problem will only worsen if they try to enhance the architecture using third-party systems that do not support ALE. If the same data set is used in different systems, it is necessary to define which is the lead system for which kind of data. For example, a call center may not be allowed to add new customers, but can correct wrong data within a data set, while new customer data sets are created and distributed in another department. The amount of data also impacts the data distribution. If we analyze the structure for RB household appliances, we see that each end-customer could have their own data set in the ERP system if third level support was provided. How can we ensure that master data is stored without redundancy and that different data sets are not created? Can we realize the necessary data consolidation beyond SAP borders and how can we include non-SAP software products? What techniques will we need to define a sophisticated distribution process for master data?

Front-End Integration

In Figure 1, we added the SAPGui to the known system architecture because this is the point of entry for SAP users at RB. The screens in SAPGui are sometimes quite crowded by elements that are not needed for every process. Furthermore, for each SAP system that you are working on, you need to open a separate window. Working on a process that affects each of the four R/3 systems requires you to open at least four windows, although everything you need could fit onto a single window. As with the other issues we have already addressed, this situation gets worse with each additional system you are using, especially if it is a non-SAP system. Perhaps you also need information from a Web site. To carry out your work tasks, you need different programs, different user names and multiple passwords! Unfortunately, most employees have only one monitor to display all these elements. Wouldn't it be nice to have only one point of access to all systems? Wouldn't it be better to show only the elements needed for a certain process?

SAP's Proposed Solution

Now that we have raised these questions, let's address some possible answers. Business systems are made up of a combination of elements based on business logic and a technical architecture. None of the systems can exist without the others, so we need to find the best way to integrate them. First, we will discuss a current concept in business logic known as service-oriented architecture (SOA) and how SAP implements it. As this strategic layer needs the technical layer as its enabler, the enterprise service architecture (ESA), we will also examine how SAP's counterpart on the technical level is realized.

Service-Oriented Architecture

Service-oriented architecture (SOA) is widely seen as the key to interoperability and flexibility requirements for future on-demand business applications. SOA supports end-to-end integration across the enterprise and among business partners. This provides a flexible business process model that allows suppliers to respond quickly to new customer requirements, new business opportunities, and competitive threats (Karch & Heilig, 2004; Zimmermann, Krogdahl, & Gee, 2004). Briefly, SOA presents the organizational view of what can be done with Web services and existing business services (Colan, 2004).

SOA is *business-oriented*, with an *integrated end-to-end process view* that focuses on *integrated data*. It is built on an *open architecture* that uses standards such as Web services and protocols such as Simple Object Access Protocol (SOAP) and Extensible Markup Language (XML).

Business-Oriented View

IT systems are expected to support business needs. This may sound self-evident, but in reality a major effort is required to support changing business needs. Therefore, SOA uses Web services for more or less complex tasks, which can be easily combined to support larger processes. In the SOA concept, legacy and newly developed systems can be integrated using Web services. To do so, legacy applications can be organized or enhanced by Web services in order to provide a well-defined interface for other applications.

Integrated End-to-End Process View

SOA does not focus on single tasks or in-house processes, but on the whole process chain from a top-level view in order to solve integration problems.

Integrated Data

The SOA concept assures data integration by handing the responsibility for data consistency and integrity over to a middleware application[ii]. This middleware can be realized using Web services.

Open Architecture

SOA does not rely on proprietary "standards;" instead, open standards are used. SOA is typically based on Web services that can also be used to enable communication between software systems. These Web services can be identified by the Uniform Resource Identifier (URI) and can exchange messages using the XML-based SOAP protocol. A central directory of available web services is usually realized using a protocol such as Universal Description, Discovery and Integration (UDDI)[iii].

SAP's Enterprise Service Architecture (ESA)

SAP developed their ESA system based on an interpretation of SOA. While SOA is a concept without concrete implementation, ESA brings the SOA concept to SAP users. ESA is a toolset that supports the definition of new business processes by using high level SAP components. These components can be implemented in various ways, but offer a uniform interface—the Web service.

In order to implement the ESA, SAP introduced a system known as "NetWeaver."

SAP NetWeaver

In order to understand the differences between the current SAP R/3 implementation at RB and a potential mySAP ERP implementation based on NetWeaver, we need to present the technical architecture of SAP R/3. Figure 3 shows how the system might have evolved beginning with SAP R/3 4.6 as it is used at RB,

Figure 3. Historical development from SAP R/3 to mySAP ERP based on SAP NetWeaver (SAP AG, 2005)

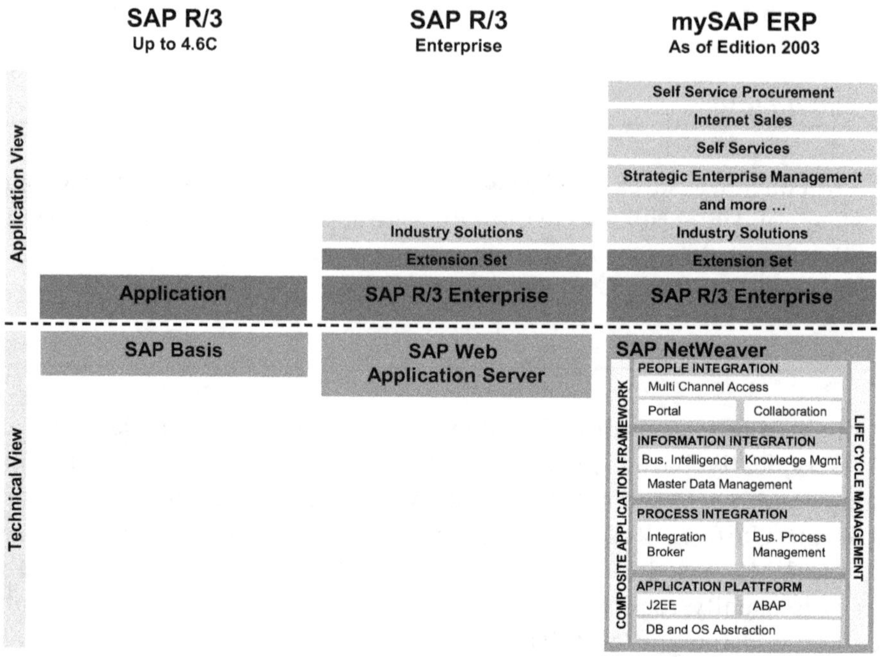

moving on to SAP R/3 4.7, and finally reaching the current version of SAP's ERP Central Components 5.0 (SAP ECC) that uses NetWeaver.

The first system shown at the left of Figure 3 contains two parts: the basis, or technical foundation, and the application logic itself, known as R/3. These elements are firmly tied together and cannot be detached or upgraded separately. This means that it is impossible to enhance technical functions without upgrading to a new release. SAP took the first step towards correcting this situation with the release of R/3 4.7, the middle system shown in Figure 3, which divided the system into the business layer R/3 Enterprise and the technical base Web Application Server (Web AS). Web AS is the foundation of all SAP products and can be upgraded independently from the business logic part of the system. Because SAP realized that there are more technical functions needed by an enterprise than Web AS alone can provide, they bundled together all the crucial technical products and optimized the interactions between them. The result is SAP's NetWeaver, the system shown on the right in Figure 3, which not only provides the technical base for a single ERP system, but for a whole enterprise. Based on the so-called "NetWeaver fridge," which contains all the technical components, business applications can be developed and run more easily. This NetWeaver fridge will be discussed further in the next section.

In an earlier section, we noted that RB would face potential issues with process, data flow, data structure and front-end integration in the future. Figure 4 shows that these elements have not disappeared, but some of them can now be found as components in the NetWeaver fridge and only the denominations of the layers differ slightly from the specified problem dimensions.

Application Platform

The basis of all SAP components is an application platform consisting of two stacks: the ABAP stack and the Java stack of the Web Application Server (Web AS). Both stacks provide an environment where runtime and development aspects are independent from the operating system and database. The Web AS Java and the Web AS ABAP stacks can function as stand-alones or as a combined solution. Both components communicate through SAP's Java Connector to allow very rapid exchanges of data. There are two reasons for the segmentation into two different stacks. The first is historical: SAP's own programming language (ABAP) has been used since the 1970s and represents a very mature base for business development. The second reason is that Java is a widely used platform, especially for Web applications, and SAP follows a best-of-breed strategy in this domain. The business applications that run on Web AS are performed at the process, information, and people integration layers.

Process Integration

The process integration layer resolves issues that arise in integrating multiple processes. The main component of the process integration layer is the SAP *Exchange Infrastructure* (XI), SAP's tool for enterprise application integration (EAI). This component supports process and data integration within heterogeneous system landscapes. It is comparable to the integration of legacy applications by the Application Link Enabling concept (ALE) within pure SAP landscapes. This NetWeaver layer addresses the identified issue of process integration, but also deals with challenges of data flow integration.

Information Integration

The information integration layer resolves many of the data structure issues by focusing on three topics: data harmonization with *master data management*, data analysis with *business intelligence*, and data organization with *knowledge management*. These topics are of great importance, especially in heterogeneous system environments where multiple data sources exist. *Master data management (MDM)* supports the implementation of a central data management, either by implementing new databases for key data sources, such as products or customers, or by defining a central access point to these distributed data sources. The *business intelligence (BI)* element of this layer represents the strategic view for business processes and is covered by its own product, SAP business information warehouse (BW). Managing the knowledge held by a business is the task of *knowledge management*, which is an integrated part of SAP's enterprise portal (EP). SAP knowledge management (KM) is the best example of how the assignment of a functional element to a layer is not product-driven, and in the next release of NetWeaver, knowledge management may even be a stand-alone product.

People Integration

The goal of this layer is to provide a single point of access for all the systems, and as such addresses front-end integration issues. The enterprise portal (EP) provides a Java-based Web platform for accessing most business systems. Another component of SAP EP allows online collaboration between users. Finally, SAP mobile infrastructure (MI) provides access to business systems by devices other than a computer, for example by using a PDA or a cellular phone (Karch & Heilig, 2004).

It is interesting to note that the analysis of future issues that we performed for RB in an earlier section closely matches the layers contained in SAP's NetWeaver. Next, we will take a detailed look at the products that are designed to implement the different functions. Life cycle management, represented by SAP Solution Manager, and the composite application framework (CAF), or xApps, will also be covered in this section as each of them corresponds directly to a product or product framework.

SAP Exchange Infrastructure (SAP XI)

SAP XI is an enterprise application integration platform that represents the communication backbone of a business landscape, providing a technical and a business view of data exchange. On the technical level, all computers can be configured with their own software components — for example, a server could be running SAP R/3 or any other software. Every software component can communicate through specific interfaces and protocols; for example, an R/3 system "talks" ALE (and its technical implementations IDoc, RFC, etc.), whereas another system may communicate only via HTTP. This connection information is also represented within XI. Knowing the message structure that can be transmitted through a specific interface is essential, as an invoice might have a different structure than a standard order. In a heterogeneous landscape, the message structure in system A may not be similar to system B — for example, message structure A might give a customer's name as "John Doe," whereas structure B stores the same name as "Doe, John." The implementation of a transformation process for this or more sophisticated problems makes a configurable mapping necessary. By combining all these elements, XI enables the technical communication between all business systems which are connected to the XI bus structure (Stumpe & Orb, 2005).

Since information specifying which software components speak which language and how we can translate the messages between components is known, it is possible to configure what the systems can talk about and who communicates with whom and when. On the business level, the messaging processes could be configured to provide direction on what to do with a specific kind of message from a system. Let's assume that we bill a customer for goods using our R/3 system A and that we transmit the invoice digitally to the customer, who uses ERP system B. This invoice transmission process starts with an IDoc (Intermediate Document, an SAP format similar to SOAP) that is sent to XI. In XI, the configuration of the invoice in the system must wait for feedback from system C before sending the message further to the customer's system B. After receiving a message from system C, XI transforms the invoice into a XML format and transmits the message via HTTP to system C. This synchronization configuration for messages enables the realization of transactional security even beyond a systems boundary (Stumpe & Orb, 2005). This is a very important prerequisite for realizing distributed processes and is needed by both SOA and ESA.

An issue we noted earlier was the problem of integrating heterogeneous systems into one process and of distributing and reusing defined processes. On the data-flow level, the problem was that of central business process monitoring and the maintenance of connections and interfaces. The process integration issues can be solved by using the different connectors provided by XI. These allow data exchange with any software system that uses a standardized interface. The distribution and adaptation of centrally designed business processes is simplified by separating the business from the technical view. A process that has been designed on the business level can be used in different configurations just by assigning the inbound and outbound ends of it to configured systems. In the solution manager section of this chapter, you will read more about how processes are designed and distributed. In addition to the central design of processes, all transactions running through SAP XI can be monitored centrally and thus provide an overview of all triggered and running communications. The central structure of this platform reduces the efforts needed to replace a system within a landscape. Instead of changing the connections to all the systems, only the interface to XI need be adjusted (Woods & Word, 2004).

SAP Master Data Management

SAP XI can solve process and data flow integration issues, but the problem of data structure integration still remains. Using an enterprise application integration (EAI) platform, all connected systems and their data can be accessed and the data changed from a central point that is connected to XI. The system that performs this function is referred to as SAP Master Data Management (MDM). MDM provides a central point for data consolidation, distribution, and maintenance. This central data point is typically constructed in three phases: consolidation, harmonization, and maintenance.

The consolidation phase provides data clearance, as all data to be managed is loaded into MDM and analyzed. The data load can be realized for different types of systems, as adapters can be used for SAP as well as for non-SAP systems. In most system landscapes, some data are redundantly stored while other data of the same type (e.g., customer data) are stored in different systems with a different structure. For example, rather than storing the entire data set for a customer redundantly, only some attributes are distributed over the entire landscape. The consolidation step performs the task of matching data of the same type in different systems. Following a detailed analysis of redundant data, data mapping beyond system boundaries can be performed to provide information about which attribute of a data set corresponds to another attribute of a data set in a different system. At the end of the consolidation phase, duplicate data and their correspondence to one other will have been identified.

Once it is known which data is stored where and how often, the next logical step (known as harmonization) is to change the data to avoid further redundant storage. Using the data mapping from the first phase, data synchronization can be established by defining which system is the lead system for which attribute of a dataset. MDM is the lead system for most attributes and data sets, as it allows a central access and overview of all data. Alternatively, specific attributes, such as the address information for a customer, can be maintained in different systems. This harmonization phase represents the implementation of the consolidation step. The result of implementing these consolidation and harmonization phases is a clean data structure within the entire system landscape.

Achieving such an exemplary data structure is laudable, but data also needs to be maintained and this represents the third phase in achieving a central data point. Depending on its use and importance, an attribute or data set can be maintained on a local system or on the central data system MDM. Data that needs to be changed quickly could be maintained in a local system, whereas important data that remains constant could be maintained centrally. As the central data administration and distribution represents an additional step in daily work, it can be supported by workflow that accelerates the process. Implementing a central data administration requires a data administrator who checks new data requests for duplicates, approves requests, and creates the needed data sets (Woods & Word, 2004).

SAP MDM addresses many of the problems that occur in larger, distributed system landscapes. Basically, redundant storage of data and thus the danger of creating inconsistency, as well as the unnecessary effort needed to type in the same data multiple times, can be avoided. Using a clean database means that business reports that are created based on this data are of higher quality and can be provided faster, since no manual adjustments are necessary. The issues that MDM covers are of great importance for both current and future topics; for example, data migration after mergers and acquisitions can be realized much faster and worldwide guidelines for data denomination and storing rules can be implemented. Such a system will shortly become critical, as vast amounts of data will be collected in the future by RFID (Radio Frequency Identification) devices. MDM thus provides a virtual data layer for the whole business landscape, which can be used for enterprise service architecture (ESA) to integrate the data provided by a wide range of Web services.

Looking back at the possible data structure issues identified earlier at RB, we see that they included the challenge of integrating non-SAP systems and the problems of data redundancy and data distribution. Using XI as a communication base, all kinds of systems could be accessed, although data harmonization requires specific adapters which will need to be implemented for all types of systems. Running through the three phases of the MDM, the redundancy- and thus inconsistency-free storage of business data throughout the landscape can be assured. As RB has already realized company-wide guidelines for the denomination and the management of specific master data, implementation of the first phase should be relatively straightforward. Implementing MDM as the central data administration point allows the design of distribution and maintenance rules that simplify the work and improve the fundamentals for decision-making. Another operational area where MDM would have been useful was when RB acquired a company called Allied Signals, mentioned earlier in the case study. The integration of the new company's data would have been easier using MDM and the databases could have been consolidated more rapidly.

SAP Business Information Warehouse (SAP BW)

We have already seen one possible role of SAP BW in the consolidation phase of the MDM introduction: the transformation of raw data into a clearly arranged overview. This process characterizes the main task of BW quite well: BW is the statistician and report-writer of NetWeaver. Such data aggregation is useful not only for technical issues, but also for business analyses. Upper-level management within a corporation requires a set of compressed and analyzed data to make informed decisions. BW refers to the software component that represents the core of the business intelligence element of NetWeaver. Like other components of NetWeaver, BW is integrated into the entire architecture and uses other elements to collect and present data. Further, it contains the main functions for data analysis and can be used as a stand-alone component. We will next study how BW interacts with other software components and other SAP products.

BW performs its tasks in three steps: data staging, data storage, and data analysis (data presentation). BW is something like a super-database; that is, it "copies" data from other systems within the landscape and aggregates it into its own database for analysis. BW gets its data from other systems using XI, as it provides access to a variety of different source systems, such as R/3 or XML data. However, the easiest and most reliable method is a direct connection to source systems such as R/3. The extracted data are cleaned within BW in order to delete redundant data sets, to correct as many data sets as possible and to detect any irregularities. The loaded data sets are then stored in InfoCubes that allow fast access to a vast array of evaluation views. The view used for the compressed data sets can be changed depending on the specific needs of the employee who is working on them. The best way to visualize so much data is to put it into pictures, which brings us to the final step in the process: the presentation of data.

In addition to classical bar and pie charts, BW provides the option of using geographical visualization, the placement of different key figures on a map. This can be used, for example, within distribution planning to identify the region with the lowest sales figures (Woods & Word, 2004). As this presentation method is realized by using dynamically created Web pages, these overviews can be integrated into the SAP Enterprise Portal presented in the next section.

However, before we discuss the SAP Enterprise Portal, let us see how RB could benefit from BW. Besides the previously mentioned option of using BW for data consolidation, together with MDM and the display of sales figures in different regions, it could also be used to identify standardization potential. For example, the productivity of plants in different countries could be examined to analyze why one plant produces a specific product quickly and another plant produces the same product more cheaply. Further application areas within higher-level business management are also possible.

SAP Enterprise Portal

BW is not the only function that can present forms and information in a format suitable for Web browsers; all NetWeaver products are Web-ready. Partially hidden Web pages are already in use in SAPGui. Thus, numerous functions and transactions can be used as Web pages or provided as Web services. Due to the underlying SAP WebAS, all SAP components can serve as a Web development platform to allow Web access to their specific features. The enormous number of Web browser and SAPGui windows would be difficult to navigate without appropriate solutions.

The SAP Enterprise Portal (EP) integrates all kinds of user interfaces, including classic SAPGui, Websites, SAP Web interfaces (Business Server Pages, Web Dynpro, Internet Transaction Server) and SAP BW Online Reports. The integrated Web site can be enhanced by Java Developments that run directly on the Web AS Java stack of EP. These representation methods can be embedded into a sophisticated authorization and display framework. This framework provides the benefit of Single Sign-On (SSO) mechanisms for all integrated user interfaces; you log on to SAP EP once and your user name and password can be mapped to your operations in SAP R/3, and into your Intranet or other software components. In addition to the pure integration, data from different interfaces or data sources can be exchanged and related by simply using a drag-and-drop mechanism called drag and relate. For example, a business overview integrated as a BW online report might show the aggregated sales figures for one customer. By dragging the sum of one customer's sales to an icon, EP displays the detailed data for all that customer's sales transactions taken from SAP R/3 or SAP ECC. SAP EP thus integrates different front-ends and different systems on the front-end side. This allows the implementation of workflows on the front-end side to go beyond system boundaries, thereby integrating all solutions represented in SAP EP (Woods & Word, 2004).

Two other functional parts that are integrated into SAP EP are the online collaboration and knowledge management functions. Online collaboration in EP allows the use of different synchronous and asynchronous communication methods. Chat, instant messaging, application sharing, and video conferencing are all supported, as are bulletin boards, team and project rooms, amongst other things. These tools provide an opportunity for personnel to interact around the clock beyond departmental and company boundaries. The provision of this infrastructure allows the joint processing of documents and makes a complex knowledge and document management system necessary. SAP EP provides its own knowledge repository or can be used with third-party document repositories to manage and distribute data. Knowledge management in EP supports versioning, document approval, subscription, rating, feedback, and discussion for any file. Large quantities of documents are an issue that must be considered in SAP EP. SAP's TREX (search and classification) has proven to be helpful in dealing with this issue. It provides an efficient search algorithm over many different document databases, as well as helping to automatically classify new files. Thus, SAP EP represents a powerful platform for Web-based applications that can be enhanced further by integrating additional functional packages provided by business packs (Woods & Word, 2004).

By comparing the identified issues at RB with the features that SAP EP offers, we can conclude that all the tasks needed by RB can be performed with this product, which makes it possible to integrate the most important front-ends into a Web-based interface and exchange data between these systems on the front-end side. Additionally, it provides users with the option to communicate in both a synchronous and asynchronous manner, as well as making it possible for them to work collaboratively on project documents.

SAP Mobile Infrastructure

It has become routine for employees to expect to obtain information on business processes from anywhere in the world if they have a notebook with a wireless LAN. However, in some situations a notebook is not an adequate device because it is not convenient enough. A PDA is sometimes a more suitable solution when a lack of space does not permit the employee to unfold his or her "mobile workplace," Both notebooks and PDAs can access Web pages, and thus SAP EP can interact with business systems. This assumes, however, that they are in some way online. As this cannot always be assured, a solution is needed that provides all necessary data and can save newly created data sets for later transmission (Woods & Word, 2004).

SAP Mobile Infrastructure (MI) provides such a solution. MI is a Java- and ABAP-based product that was designed to solve exactly these problems. It consists of a server component, centrally installed, and a Java-based client component that is installed on the mobile device, such as the notebook or PDA. The server part can interact with SAP and non-SAP systems, accessing or modifying the correspondent data. A sales representative, for example, can download necessary data from an ERP or CRM system to his mobile device, visit a customer, type in new standard orders that are stored locally, and transmit them later to the central system when he reconnects. This is one of the scenarios that is supported out of the box. Other applications can be developed centrally and distributed to mobile devices (Woods & Word, 2004).

We did not address the issues covered by SAP MI earlier, as they were not mentioned in the case study. However, off-line access to important data could be an important issue at RB. Even if the described sales scenario is not vitally important for RB, as they have a predominantly business-to-business structure, the following examples will point up the role of MI. Imagine a warehouse employee who is making a clearing posting: He goes to a storage rack to check out the exact denomination or number of the work-in-progress, makes an entry in a pad, and than goes back to his computer to type in the necessary data later. Now imagine the same transaction using a PDA with a barcode or a RFID scanner attached: He goes once through the warehouse, scans all goods to be moved and then goes back to his work station to synchronize the data. Similar behavior can be observed by technicians working in the field who order spare parts. Normally it would take days while forms are completed and submitted to order a part. On a decision-making level, MI can provide the data needed to make an off-line business analysis, to make decisions, and to transmit them later to the central system. Thus, MI has many application possibilities in companies such as RB.

SAP Solution Manager

Now that we have discussed the middle column of the NetWeaver fridge shown in Figure 4, we will deal with the two cross section components, xApps and Solution Manager (SM). SM will be examined first. This component supports the complete life cycle management of business systems, from the installation, adaptation, monitoring, technical and process changes, to the obsolescence of a product. SM centralizes a conglomeration of many functions within a system landscape. The range of functions can be divided into two groups: technical functions and business functions. The technical functions include central monitoring of servers and software and processes that allow the definition of service-level agreements. In addition, a helpdesk for all connected systems and a knowledge base for solved issues are included (Woods & Word, 2004).

Although all these functions are very important for a large business system, we will focus on the business aspects of functionality. The business view in SM includes tools that define and distribute processes and define business system implementation guidelines.

SAP systems are usually delivered as plain vanilla — that is, without settings — and are thus unusable in a business sense. This is where consultants enter the picture: They adapt the business system to meet a company's specific needs.

Figure 4. SAP NetWeaver overview (Karch & Heilig, 2004)

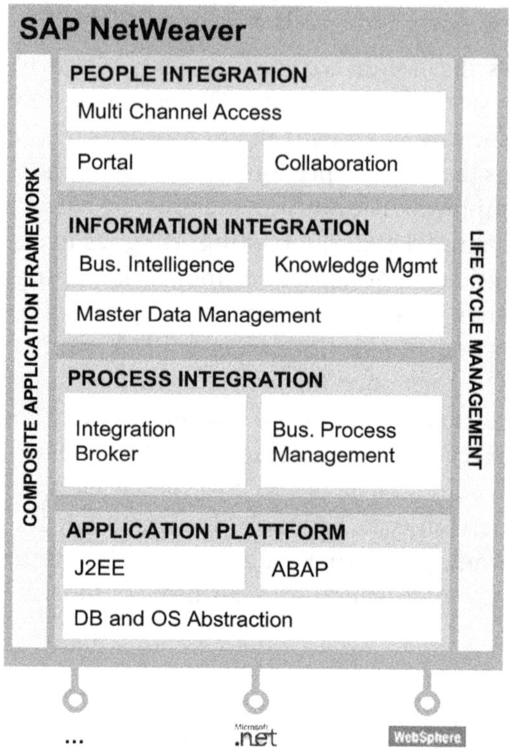

This adaptation normally takes several months of manpower as the existing state of processes and organizations must first be analyzed, then improved, and finally implemented. To accelerate and simplify this complex task, SAP has developed a set of guidelines which contain the most common steps involved in this customization and has integrated it into SM in the form of a "business roadmap." These guidelines contain descriptions of the steps to be taken, shortcuts to relevant transactions within an SAP system, and ready-to-use documents such as meeting announcements and default letters for different purposes. For every type of system, there is a specific roadmap that structures tasks. For example, SAP R/3 has guidelines that differ from those for SAP CRM because the necessary settings are not the same. These roadmaps can be modified depending on the specific company needs. The adopted roadmap can be transported into the system to be customized. At this point, each consultant receives his own little task packet (Woods & Word, 2004). Customizing projects consists to a large extent of facilitating communication between the end-users and consultants and performing recurrent tasks, and the task packets provided by SM are very helpful in this process.

Developing a new system to maintain the status quo is one thing; however, during the lifetime of a business system, processes change and need to be redistributed across the landscape. Accommodating a changing business environment represents the second function group of SM's business view. With the use of an additional tool called ARIS for NetWeaver, business processes can be modeled and implemented in one or more systems. The advantage of this method is that the processes do not need to be programmed, but rather designed. They are transformed into their technical implementation automatically using SM and XI. This automation requires the use and proper configuration of all the relevant components. ARIS for NetWeaver is a design workbench that provides a pure business view of processes. The business processes are exchanged with SM, where they are divided into a series of small steps and customization activities. SM is delivered with its own repository of reference processes that can be exchanged with ARIS, modified, and sent back. Thus, SM serves as a repository for the specific processes within a corporation. Processes that are already split into steps can be implemented by synchronization with XI. Using the EAI platform allows for the creation of cross-component processes that reach from SAP to non-SAP systems (Karch & Heilig, 2004). Implementing such a system automatically is in reality more sophisticated than we have described here, but this description should give you an impression of the possibilities for process design using NetWeaver. SM, together with ARIS, provides a more effective way to elaborate and implement business processes.

At RB, there was an issue concerning the design and distribution processes throughout the plants: SM could be used to simplify this. Processes would be designed in ARIS, transmitted to SM, adopted to the specific needs of each plant, and finally distributed by implementing them in either a central or in several distributed XI systems.

SAP xApps

SAP's xApps provide the link between the technical components of NetWeaver and the concept of SOA. The entire technical infrastructure that NetWeaver provides is aimed at integrating all business systems and providing the foundation for easily designing cross-component application — Composite Applications (xApps), which run in SAP EP.

The smallest unit in SOA is a Web service. SAP calls Web services that are supplied by business systems "application services." These services provide a small element of functionality that can be combined with others to create a logical unit around a specific business object. For example, all services for managing invoices, such as creating, modifying, viewing, and so forth, can be consolidated into an enterprise service. In object-oriented programming terminology, the enterprise service, will be considered as a class and the application or Web services will be a method of implementation for this class. Using these classes and methods, new applications can be built that implement processes spreading over all systems connected to the NetWeaver infrastructure (Karch & Heilig, 2004).

The hierarchical structure used to organize the services that already exist within the landscape is part of the component application framework (CAF). The idea behind CAF is to deliver a new kind of application, the packaged xApps, which uses the features of the NetWeaver components. NetWeaver is "only" the technical infrastructure for process-oriented composite applications according to SOA (Woods & Word, 2004). Thus implementing SOA or ESA requires a change from a functional view to the type of process-oriented thinking represented by a fully integrated platform for business applications.

The interaction of all NetWeaver components is visualized in Figure 5, where different business systems are embedded in the infrastructure provided by SAP's platform. When this is compared to Figure 1, the figures show that implementing Netweaver at RB will integrate all the systems on the application level. The applications can then exchange data using the EAI platform XI, and

Figure 5. Interaction of the NetWeaver components with SAP and non-SAP business applications

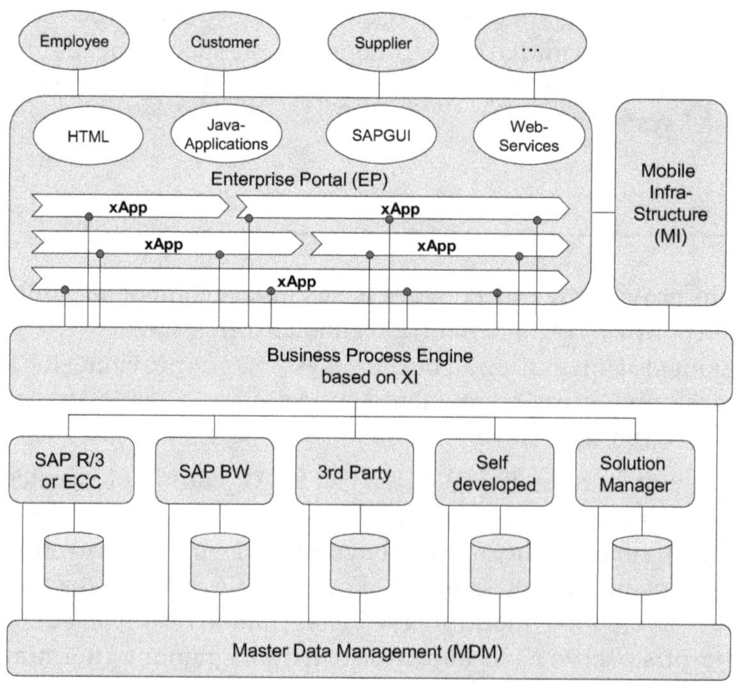

data synchronization and distribution is realized using MDM. MDM loads data directly from all the business systems, but manages the harmonization through XI. Although BW and SM play an important role, from a technical point of view they are application systems and provide information analysis and process management functions.

Most application services can be accessed directly by addressing a specific business system, but all services and connection possibilities should be managed by XI as a service repository. Thus, XI represents the communication backbone for the landscape that enables all the systems to talk to each other. Additionally, XI synchronizes all the messages — that is, only if all the messages belonging to a transaction arrive in XI will the transaction be completed and the databases changed. All services registered at XI can be accessed directly by the front-end layer (EP or MI), but can also be used by xApps running in EP. Furthermore, Web pages, Java applications and SAPGui can all be used within EP. The four issues identified at the beginning of this chapter are all resolved by using this architecture. MDM represents a virtual level that provides an access method for all the data in a business landscape. As a result, all xApps use the same database. Process and data-flow integration are represented by XI and SM, data structure integration by MDM (and XI) and, finally, front-end integration is performed by EP and MI.

Migration from a 3-Tier Architecture to SOA

The shift from a *3-tier* architecture to an *SOA* or, more specifically, SAP's SOA implementation ESA as outlined in Figure 6, is just as revolutionary as was the earlier shift from a mainframe architecture to a Client/Server 3-tier architecture.

Figure 6. Movement toward a SOA from a SAP perspective (Karch & Heilig, 2004)

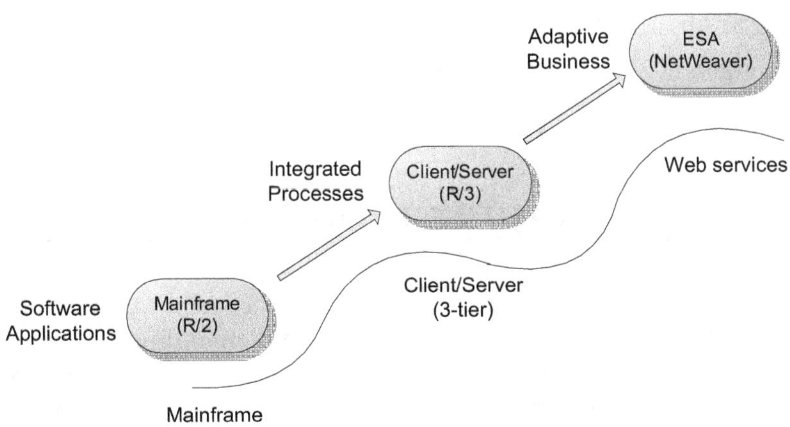

The shift to ESA is revolutionary because it demonstrates how the implementation of IT-Systems can be put into practice in a business-driven manner. For example, implementing a CRM-System at RB could be looked at from a functional or internal process view and isolated functions such as a customer data update can be performed. Alternatively, the system can be viewed from a business service perspective, and different customer services such as customer information campaigns, call back actions, and accounts receivable management performed. This concept has the best potential for satisfying the customers and fulfilling a company's near- and long-term needs.

But why is this shift revolutionary? Could this not have been done previously with the 3-tier client/server architecture? From a technical point of view, it is possible to implement the same functions using the established and well-proven client/server architecture. However, one needs to possess both in-depth business knowledge and in-depth technical knowledge in order to complete the task successfully.

The revolutionary aspect of service-oriented architectures is that they enable decision-makers to formulate their business concepts as business services in a way that allows them to implement the concepts directly as services in a service-oriented IT system. The revolution can be better visualized in terms of building a house from prefabricated parts which can be flexibly combined, instead of using individual raw materials to build the house. The time savings, flexibility, and reduced need for skilled labor are enormous with prefabricated construction. Similarly, the use of NetWeaver will lead to a much better alignment of business needs and IT applications.

The implementation of an enterprise service architecture is less of an implementation than a migration. There are several reasons for this:

- Firms already have fine-tuned client/server environments which fit the actual tasks (Wittges, Mohr, Krcmar, & Klosterberg, 2005)
- The ESA architecture is not an out-of-the-box solution (Agassi, 2005)
- ESA and NetWeaver are still under development

Chapter Summary

In this chapter, we discussed the future directions in SAP R/3, in particular the service-oriented architecture (SOA) and how implementing this architecture might help RB solve its process, data flow, data structure, and front-end integration issues. We also showed in this chapter why it is important for you to be familiar with both the well-established client/server-based solutions, which are currently up and running, and the ESA-based solutions that are on the way, so that you can offer firms an improved value proposition when you come to look for a job.

Team Assignments and Discussion Questions

1. What are the (integration) layers of ESA?
2. Summarize the tasks and the importance of the front-end layer within ESA.
3. Summarize the tasks and the importance of the data structure (MDM), process integration (XI), business intelligence (BW), and process design (SM) layers.
4. What general benefits can be achieved by using ESA?
5. What are the prerequisites for the implementation of ESA and NetWeaver?
6. Can you imagine the implementation of a NetWeaver component outside of the ESA concept? If so, how could it be performed?
7. Discuss the pros and cons of leading an SOA implementation from a business and from a technical point of view.
8. What is the role of the data structure and the process integration layer for the realization of the transaction concept?
9. What issues arise when you try to string several incoherent services together?
10. What influence could NetWeaver have had on the implementation process at RB if it were available at that time?
11. Compare the implementation of a business process in the current RB infrastructure and in ESA.
12. Plan the steps needed to implement ESA in the current infrastructure at RB. Which dependencies have to be taken into account?
13. Discuss the pros and cons of a top-down and bottom-up implementation process for SOA.
14. Compare the change from a client/server architecture to SOA to the change from a functional to a process-oriented business organization. What conclusions can be drawn from this comparison?

References

Agassi, S. (2005). *Weblog*. Retrieved July 1, 2005, from https://www.sdn.sap.com/sdn/weblogs.sdn?blog=/pub/u/6806

Carr, N. G. (2003). IT doesn't matter. *Harvard Business Review, 5,* 41-49.

Colan, M. (2004). *Service-oriented architecture expands the vision of Web services, Part 1.* Retrieved September 7, 2005, from http://www-128.ibm.com/developerworks/library/ws-soaintro.html

Karch, S., & Heilig, L. (2004). *SAP press – SAP NetWeaver*. Bonn, Germany: Galileo Press.

Krcmar, H. (2005). *Informationsmanagement* (4th ed.). Berlin & Heidelberg, Germany: Springer.

SAP AG. (2005). *ESA discussion SAP developer network*. Retrieved June 20, 2005, from https://www.sdn.sap.com/sdn/esa.sdn?node=linkDnode12

Stumpe, J., & Orb, J. (2005). *SAP press - SAP exchange infrastructure*. Bonn, Germany: Galileo Press.

Wittges, H., Mohr, M., Krcmar, H., & Klosterberg, M. (2005). *SAP®-strategie 2005 – Die sicht der IT-entscheider* (Study). Munich, Germany: Technische Universität München.

Woods, D., & Word, J. (2004). *SAP NetWeaver for dummies*. Hoboken, NJ: Wiley Publishing.

Zimmermann, O., Krogdahl, P., & Gee, C. (2004). *Elements of service-oriented analyses and design*. Retrieved September 7, 2005, from http://www-106.ibm.com/developerworks/webservices/library/ws-soad1/

Endnotes

i The materials in this chapter were contributed by Valentin Nicolescu, Holger Wittges, and Helmut Krcmar of the Technische Universität, München

ii Middleware is connectivity software that consists of a set of enabling services that allows multiple processes running on one or more machines to interact across a network.

iii The Universal Description, Discovery, and Integration (UDDI) protocol is one of the major building blocks required for successful Web services.

UDDI creates a standard interoperable platform that enables companies and applications to quickly, easily, and dynamically find and use Web services over the Internet. UDDI also allows operational registries to be maintained for different purposes in different contexts. UDDI is a cross-industry effort driven by major platform and software providers, as well as marketplace operators and e-business leaders within the OASIS standards consortium. For more details, consult http://www.uddi.org/.

Chapter XI

Another Example of Use of ERP Systems in a Multinational Company

Learning Outcomes

- Discuss how an ERP system was implemented at Sidler GmbH
- Address the decision-making process in selecting SAP as a company's new ERP system, and managing continuing problems with its implementation
- List lessons in implementing ERP systems by studying this case study

Introduction

We have so far discussed the Robert Bosch case study and showed how the company's IT systems evolved so that it has standardized on use of SAP R/3. In this chapter, we provide another example of a company that has implemented SAP R/3 systems. The case study is about implementation of SAP R/3 at Sidler GmbH, a leading supplier of interior lighting and trim products for automobiles. The case addresses the decision-making process in selecting SAP as the company's ERP systems and the continuing problems with its implementation.

Sidler GmbH & Co. Case Study[i]

Light means orientation and safety. We give you the best possible lighting in today's most popular means of transportation: the automobile.

Sidler Brochure

Sidler was a leading European supplier of interior lighting and trim products for automobiles. It was 1997 when Dr. Helmut Rapp, Sidler's managing director, decided to address the challenges of operating under Europe's new common currency, the Euro, addressing the millennium's Y2K problem, and upgrading its information systems to deal with the company's rapid global growth. Since joining Sidler in 1994, Rapp had successfully grown the company from 85 million deutschmarks (DM) to over 210 million DMs in 1998. Sidler's existing data processing system no longer provided management with the information needed to determine the profitability of its products, or timely data from its geographically disbursed operations. In his attempt to minimize disruption to his organization, Rapp had continued working with R&H (Rembold & Holzer, Breisach; now Brain AG, http://www.brain-ag.com), its existing software supplier. Unfortunately, R&H had been unable to provide assurances that its new software version would meet Sidler's needs.

Sidler had relied on R&H for its key EDP software needs. But R&H had not developed new software to address the needs of the Euro, or to address the growing complexity of global operations. They worked on a new release but exact details were not available. Since Sidler also had a limited EDP staff, they had relied on outside vendors for their software needs. Mr. Kaltenbach, the primary EDP professional at Sidler, was skilled at keeping the company's AS/400 computer running R&H's systems on a day-to-day basis, but to address Sidler's future needs, Dr. Rapp asked a team, headed by Kaltenbach, to find a solution to Sidler's needs in the new millennium. Sidler felt that its current AS/400 computer system could continue to provide the platform for the future operating system. Firms like PSI (PSI AG, Berlin, http://www.psi.de), a respected ERP supplier, SAP, and R&H, Sidler's current supplier, had software systems that would run on their existing platform.

A visit with IBM, a key SAP vendor, did not go well. In contrast, a visit with SLIGOS-Industries (now called Atos Engine, see http://www.atos-group.de) went quite well. In fact, the SLIGOS R/3 Automotive Supplier software (Appendix 1) had adapted SAP for the automotive supplier industry. In addition, SLIGOS was one of the major SAP consultants in Europe and promised to provide adequate support for the implementation of SAP.

SLIGOS also arranged for Sidler team members to visit a number of SAP users. Sidler's management board now had to decide on the best course of action.

Sidler's Management History

Since 1925, Sidler had been a family-owned business that produced all types of metal work. In 1951, Sidler's owners began making decorative chrome for automobile headlights and windows. As plastics replaced chrome trim, Sidler developed simulated wood trim products and entered the interior lighting business. By the early 1990s, when family members wanted to move out of active management, Dr. Rapp joined Sidler. At that time, Sidler had annual revenues of DM 85 million. Dr. Rapp explained:

I found that Sidler was looking for someone to build a non-family management. I liked the business and joined the company in early 1994. We have more than doubled the size and have gone international. It was not very dynamic when I joined, and it was necessary to push forward or fall backwards. Sidler had manufacturing processes in two-color molding that nobody could copy, but they needed to move more quickly. They have learned to do that.

Dr. Rapp had earned his PhD in engineering in 1980 at Stuttgart University. He explained his experiences prior to joining Sidler:

I worked with Rochester University to develop a 3D CAD system and then joined the consulting arm of the Ingersol Milling Machine Company in Rochford, Illinois called Ingersol Engineers. It was a good education for five or six years, but you never stay home. It was interesting, but hard on the family. I was officially based in Düsseldorf, but was never there.

As a consultant, you never get to take something to the end. When I decided to leave consulting, Krupp gave me an offer. They had just bought a company in Stuttgart and hired me as the assistant to the new CEO. I stayed until my division was a DM 250 million business, but because of the politics, I decided to go into small businesses.

My first firm was nearly bankrupt, with debt equal to sales. They could not pay it off, but had a market and product that was needed to build cars. They did things that nobody else could do. That was challenging. A team of three turned it around. I did the consolidation of manufacturing and reduced the headcount from 1600 to 900 with positive cash flows in nine months. We could then get new equity and bank support. It was then sold to a U.S. firm, but I didn't want to work for them.

The Sidler Group

In 1995, Sidler became a group structure after acquiring a firm in the U.S. It was in the same business as Sidler, but the cultural differences caused some communication problems. That acquisition made Sidler a small global player, with operations on either side of the Atlantic Ocean. The Sidler group now includes three locations in Germany around Tübingen called Sidler, Sico, and Smig. Sidler had two Rivoret factories in France, plus a Sidler Apag R&D and sales department in Paris. Apag was its Swiss electronics company. Sidler Ltd. was a new factory located close to Birmingham, England. It also had three factories in the U.S. in or near Detroit: Sidler Inc. and Sidler Plastics, and the Laotto Division in Indiana. Sidler had a new facility in Brazil, Sidler Ltd., in Sao Paulo. The company was working to establish operations in Portugal, the Czech Republic, and South Korea. (See Appendix 2 for Group information.)

By 1998, Sidler had become a DM 200 million supplier to the automotive industry (see Figure 1 for company information). Dr. Rapp explained the changing role of the headquarters operation:

We have had substantial growth rate though our projections are always low. We try to keep projections low. Last year we had group revenues of over DM 200 million. Headquarters' revenues are declining as we make it a marketing and engineering support center. We also do the development and testing of production processes here. It is a big playground for engineers. Of 16 machines located here, only two are doing production and the others are for testing. The last four months of last year, we started an annual business volume of DM 50 million that included over 200 tools to receive, to test, to optimize, and to ship to France and England for normal production.

Sidler's Business

Sidler was the market leader in interior lighting, ashtrays, and other interior trim products for cars. Rapp's goal was to be a world-class supplier in its business niche, and to have open cooperation with its customers in R&D and production.

Figure 1. 1998 information for Sidler GmbH & Co.

Group Sales: 210.3 million DM
Sales in Lighting Products: 127.6 million DM
Sales by Product Group: 63.4% lighting, 17.6% interior trim, 19% electrical/electronics
Sales by Region: 55% Germany, 25% rest of Europe, 20% rest of world.

Sidler Strategy: Become a total global player in lighting and interior trim while providing complete customer service. Sidler will supply all carmakers to spread the risk of dependence on any single customer or region. Sidler has five engineers charged with developing 12 new patents per year. Key technologies include Ruby Optics[1)] (low voltage with low light), rail light systems[2)], clip and fit lamps[3)], and behind the surface lighting[4)].

Group Companies

- Sidler GmbH & Co., Tubingen, Germany
- SMIG GmbH & Co., Reutlingen, Germany
- GHS Stanz+Montage GmbH, Riederich, Germany
- Rivoret SaRL, Brumath at Garches, France
- Sidler APAG SaRL, Haguenau at Garches, France
- APAG AG, Dubendorf, Switzerland
- Sidler Ltd., Telford, UK
- Sidler Inc, Madison heights, Michigan, USA
- Sidler Plastics Inc., Rochester Hills, Michigan, USA
- Laotto Manufacturing, Laotto, Indiana, USA
- Sidler Ltda., Sao Paulo, Brazil.

Sales by Customer

24.4% DaimlerChrysler	6.7% Ford
16.5% Audi/VW	6.6% Volvo
10.3% BMW/Rover	4.3% Toyota
8% Renault	3.5% Sogendac

[1)] A special optic for LEDs which we use in high market stop lamps with the effect to reduce the number of necessary lights
[2)] Used in light pick-up-trucks where we snap lights etc. on rails at the roof so they could be moved as required.
[3)] Special design of cheap interior light.
[4)] Fakewood on transparent material with signal lights behind which is used in dashboards with the effect to reduce necessary space in the car.

Rapp explained:

Sidler is a leader in interior lighting in automobiles. Small lights and lamps account for 70.1%[ii] of our business. That includes interior lights, comfort lights, small indicator lamps, stop lamps, and license plate lamps. We are rated as everybody's "best of."

Interior trim accounts for 28.6% of our business, including ashtrays, coin boxes, deposit boxes, and cup holders. For example, we supply Mercedes interior trim includes ashtrays, cup holders, and coin boxes. Other business includes fuse boxes, motor compartment, off delay units, and dampers. We do some of the electronics for interior lighting in Switzerland. We do some innovative things like integrated headliners with lighting and boxes and neon lights.

Figure 2. Sidler's global market position

Market	Competitors	Sales	Market share
France	Sidler AXO[1)] SLI[2)] Other	20 Million DM	55% 25% 15% 5%
Europe	Sidler Hella[3)] SLI OLSA[4)] AXO Other	150 Million DM	35% 25% 10% 10% 5% 15%
USA	Sidler Prince[5)] Donnelly[5)] Hella Other	20 Million DM	10% 35% 25% 10% 20%
Japan	I.I. Stanley Other		75% 25%

[1)] family own business in France [2)] small french competitor
[3)] see http://www.hella.de [4)] an italian competitor
[5)] Prince and the interior lighting of the lighting division of Donnelly has been bought by Johnson Controls, see: http://www.johnsoncontrols.com/.

Sidler's Customer Base

Sidler was the market leader in Europe, with roughly 35% of the interior lighting market, and roughly 12% of the world market (Figures 2 and 3). Sidler had been successful in continuing to expand its market share. Rapp explained:

Our customer base has been with us a long time. It includes everybody who builds cars. Our share of sales is very balanced, except for Mercedes Benz since they are located very close to us. Normally we try to keep any single customer below 25%, since I want to be able to walk away should they start asking nasty questions. That happened with a Mr. Lopez who told us what the prices should be. We stopped working with VW for some time, but we are back in now.

Having a wide range of customers is our form of diversification. We are only in the automobile business, and don't sell product for trucks or busses.

Figure 3. Sidler's customer base

Producer	Supplier	Market Share
Renault	Sidler AXO Other	65% 30% 5%
Fiat	OLSA AXO/SEIMA Sidler	90% 10% Introduction
VW/Audi	Sidler SLI FER[1)] Other	30% 25% 25% 20%
Ford	Donnelly Hella Other Sidler	30% 25% 25% 20%
Opel	Sidler Hella Other	50% 25% 15%
BMW/Rover	Sidler Hella Eaton[2)] AXO/SEIMA Other	35% 35% 10% 10% 10%
Mercedes	Sidler Hella	50% 50%
Japanese Transplants in Europe (Nissan, Toyota, Honda)	Sidler Hella Other	65% 30% 5%

[1)] joint company of BMW and VW as an go east initiative
[2)] an original manufacturer of switches

We know how much our customers depend on quality, delivery, and total value; on zero-defect manufacturing and sound, economical logistics even within this small market segment. In quality, we have QS9000/VDA 6.1 and ISO 9001 through the entire production chain.

Sidler provided interior and trim product (Figure 4). Sidler's strategy was to provide excellent service to its customers. Rapp explained:

We sell to everyone in automotive and try to speak everyone's language, in terms of EDI protocols and CAD systems. We operate different CAD systems in different countries, but they are all on one network. We use Catia, UniGraphics, Euclid, PDGS or whatever CAD systems the major manufacturers use. We have established permanent customer teams to stay in contact with customers.

Figure 4. Sidler product line

INTERIOR Dome Lights Reading Lights Cargo Lights Courtesy Lamps Puddle Lamps Door Reflectors LED Illuminators Luggage Lamps Glove Box Lamps Overhead Console with Electronics	**EXTERIOR** License Plates Indicator Lamps Side Repeaters Side Markers Reflex Reflectors
TRIM Interior Trim Exterior Trim Appliques Bezels Manual Ashtrays Damper Ashtrays Deposit Boxes Coin Boxes Cupholders Radio Flaps Consoles	**SURFACE DECORATION** Common Paint Soft Paint Phote-Repossess Design Molded Color/Texture Vacuum Metalizing

But the goal is to provide the best in quality, price, and service. Our interior lighting and ashtrays meet the highest standards of quality, ergonomics, and design. We integrate complex electronic systems and storage compartments into dashboards and consoles that please the eye and are simple to use. We are happy when the driver does not notice our products, though our products add to their comfort.

Industry Evolution

Sidler's business focused on the interior parts of an automobile that are visible to the customer who buys the car. As global automobile competition increased, interior features were becoming increasingly important in the car-buyer's decision to purchase new cars. While some people were limited to buying the lowest-priced car, the aging baby boomers wanted much more.

For example, Audis were gaining popularity in Europe. Baby boomers wanted more comfort and safety. Rapp explained the trend:

Since people spend more time in their cars, luxury lighting, fresh air, and more functions are gaining importance. In an automobile, light and air have to be right all the time. Everything should function smoothly without having to consult a bulky owner's manual.

As a result of changing driver needs, the vehicle manufacturers have to address these issues. They are trying to outsource more module-type business to their suppliers. For example, in 1994, we supplied 100% to OEMs. Today, 30% to 40% of our business goes to large first-tier suppliers. For example, all of the interior of the BMW Z3 that is produced in Spartanburg, South Carolina, is supplied by Lumiere Allibert, a French firm. They had not worked in the U.S. before, and needed suppliers that were local there and knew the market. They also wanted suppliers that knew BMW design engineers and produced in the U.S. In the area of interior lighting, we are the only firm with sales, R&D, and production in both the U.S. and EU. Large firms like Donnely (see http://www.donnelly.com) are centralized in the U.S. If there is a quality problem or design change in Europe, Donnely must go back to the U.S.

In 1998, there were over 350 automotive suppliers that changed ownership. It was becoming difficult to know who owns suppliers today. Sidler was involved in a SAAB 9000 project in which its business partner changed names four times over the project's life. It now belongs to Johnson Controls. Delphi, the spin-off from General Motors, belongs now to Johnson Controls.

For Sidler, the market was huge. As manufacturers moved to platform strategies to reduce costs, an increasing number of parts were identical between different models of the platform. For example, all of the models developed from BMW's new 3-series were identical up to the dashboard. Automakers are also trying to introduce more niche cars, including speedsters and roadsters, or MPVs. Mercedes modified its C-class to make the CLK coupe in order to reach more niche markets. While the exterior might change slightly, the interior would change a lot. Rapp explained:

Mercedes added more fake wood or chrome. You need another handy holder, cup holder, or gimmick that you can push. They don't make sense, but people want them. We make those nice things. Every face-lift gives us a new business opportunity. We don't have to wait for the next generation platform, because our products are not under the hood. Light green is very popular in Italy right now, so we make a lot of parts that color. We are only limited by our ability to swallow growth. We are trying to slow growth down.

Figure 5. Lighting requirements by vehicle class organization

Vehicle Class	Installed Wattage	Average Wattage	Price in DM
Small cars	5-30 W	16.1 W	3.30- 20.00
Mid-sized class	15-100 W	34.1 W	10.00- 40.00
Large-size class	20-150 W	57.6 W	33.00-667.00
Large Vans	20-50 W	41.5 W	16.00- 60.00

We have grown 20 to 25% per year. That stretches people and capital. You have to follow it with the EDP systems. E-mail is standard here, but consolidating our parts cost data seems to be very difficult. You need systems in place, and we are working on it.

The need for technical solutions to design was also increasing. Technical functions were increasingly being included in the car's lighting components. For example, the BMW internal lighting system stays on and slowly dims after the passengers leave the car. The lights also come on when the remote key is pressed to show the driver the car's location and make it easy to see inside the car at night. Design aspects are also becoming more important as consumers realize the importance of lighting for the internal décor and function. Rapp explained:

We have developed simulated woodtrim that includes lighting behind it. New developments have new economic consequences to customers, like "clip and fit" small lamps, and special optical effects to reduce voltage requirements, but not light levels. Finally, there is a trend to reduce the total number of parts used in products. Today, the value added from purchasing is around 60% for automobiles.

The amount of lighting and products also differed by vehicle size. Larger vehicles often had up to three times the requirements of small cars, as shown in Figure 5.

Sidler was functionally organized into sales, logistics, quality, technical (R&D, design, tooling, and production), and finance departments. But Rapp used teams to overcome the formal structure. He explained:

In real life, we operate a matrix organization. Since nobody understands the concept of a matrix organization, we have established customer sales teams that include design engineers, logistics people, and quality people. They report to their functional area, but are located in the customer team area.

Figure 6. Sidler's EDP system in 1997

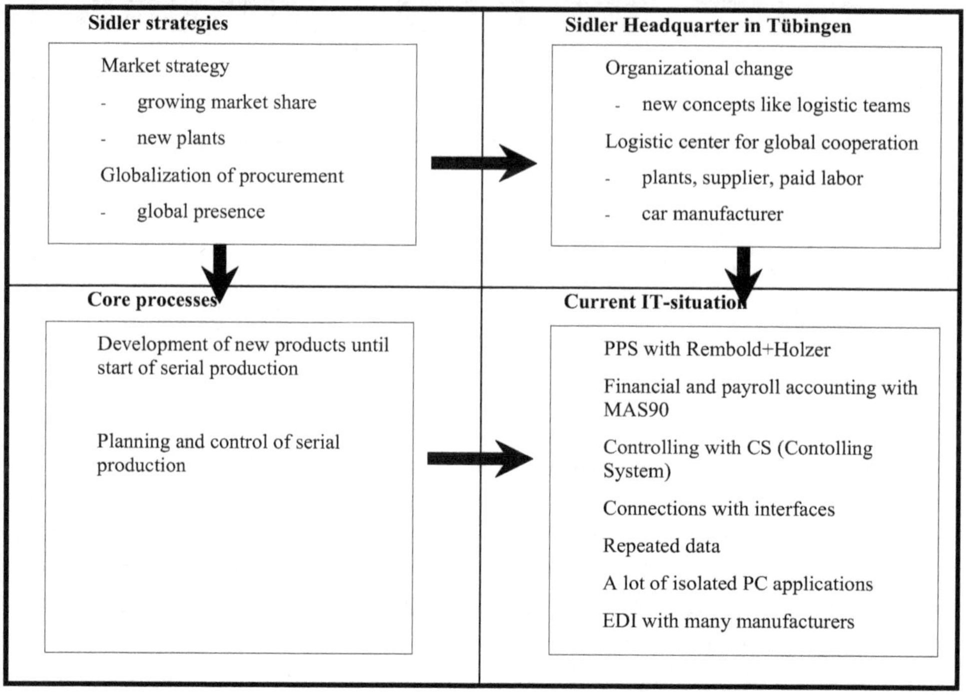

Sidler's EDP Challenge

When Rapp joined Sidler in 1994, it had one computer system to run its operational systems, including PPS[iii], purchasing, and sales and distribution. A second computer was used for accounting. The control people who compiled the financial results used a third computer. To some extent, having fragmented systems was the philosophy of the family owners, who preferred the security of not putting all information together. At the end of each month, the owners would look at the three reports and, if they were close, considered it all right. One of Rapp's first challenges was to improve the EDP capabilities of Sidler. He explained:

We had problems handling growth, and we had hardware that needed to be replaced. So in 1994, we went to one computer system with a higher degree of integration on one platform. It was all serviced from outside and we had no internal EDP group. Even if the machine was down, we had somebody from outside fix the problem. When a law changed in payroll, we called

Figure 7. Current situation in different application areas

Actual Situation	Weaknesses	Potential of improvement by R/3[1)]
Human Resource (HR) - payroll accounting with MAS90 - time-management with R+H	- different systems, no integration, repeated master data - reliance on two vendors - manual efforts for external statistics	❸
Sales and Distribution (SD) - scheduling agreement processing with R+H - no tool-handling - delivery schedule processing via EDI - 150 kinds of containers for distribution/shipping - self billing	- no commission settlement - no container oriented inventory management - repack is necessary - bad overview about returnable transport packaging (own/provided by customer) - manual import/export reporting - manual effort self billing verification - manual effort for custom's service processing	❶
Material Management (MM) - normal purchase orders - master agreements	- quality management not integrated - no vendor evaluation according the German Automobile Manufacturer Association - complicated flow of provided material for subcontractors	❸
Production Planning and Control (PP) - material master data, bills of material, routings and work centers are processed with XPPS of R+H - variants are handled with single level BOMs - PC-based management of production resources and tools (PRT) - MRP is based on forecast delivery schedule	- Support of alternative production is bad - validities are not supported - inconsistent data - no overview about the availability of PRTs	❷
Management of development projects (PS) - basic dates planning with MS-Project	- no current planning - no overview about budget and actual costs - no cross company reporting - no standard work breakdown and procedure	❸
Finance accounting (FI) - A/R and A/P with MAS90	- high manual effort in AR - no proposal list for open items	❸
Controlling (CO) - CS-Software is flexible and is well accepted by users (http://www.cs-controlling.de)	- a lot of effort is necessary to get the data via interfaces - high effort for primary cost planning per cost center	❷
General	- bad documentation and help facitities - inflexible printer allocation - bad service of the vendor - bad training	❸

Rating Scale: 3 high, 2 middle, 1 low

somebody to change the parameters in the program and we fed in the changes.

With such a weak EDP knowledge base, I couldn't change the MRP system. We kept the old MRP system and linked it with an IBM MAS 90 module for accounting and CS Control module[iv]*. That was our first EDP project. We*

Figure 8. The EDI-support is a key success factor in the operation of a component supplier

EDI Partner	Forecast delivery schedule	JIT delivery schedule	Shipping notification	Billing and self-billing	Forecast delivery schedule with supplier	SAP R/3
DaimlerChrysler	✓		✓	✓		✓
BMW	✓	✓	✓	✓		✓
Ford	✓	✓	✓			✓
Porsche	✓		✓			✓
VW/AUDI	✓	✓	✓			✓
Opel	✓		✓			✓
Nissan	✓		✓	✓		Planned
Honda	✓		✓	✓		Planned
Peugeot	✓		✓	Planned		Planned
Renault	✓		✓			Planned
Saab	✓		✓			Planned
Volvo	✓		✓			Planned
Rover	✓		✓	✓		Planned
SMIG					✓	Planned

brought in some people to do that and kept them so we could build our expertise. We lived with that situation until the beginning of 1997.

In early 1997, we began discussing the Euro and Y2K issues. Our partners for the IBM MAS 90 system told us that IBM was planning on dropping that product. Its support would be outsourced. While the manufacturing modules were also supposed to be upgraded to handle the Euro and Y2K problems, our vendors didn't know exactly when.

In early 1997, Dr. Rapp put together a core team of three plus five outside experts to evaluate Sidler's options. They described the existing system and software as shown in Figure 6. They then looked at possible solutions on the market, such as SAP and Baan. Rapp explained his strategy:

I wanted those working with the systems to tell me what they wanted. If I made the decision, they would tell me every day why it was a bad decision. They evaluated several new systems while looking at the future release of our current PPS system. We discovered that the new PPS release was totally new, including the user interface and hardware platform. When we saw the cost of the upgrade, *we felt that it was a new system and felt that this was the right time to consider a change, since nobody liked the current system anyway.*

Figure 9. Aid to decision-making for future data-processing systems

1. Comparison of core problems

	R&H Release 1.6	SAP R/3
Data exchange transaction with Edifact (required by Opel)	Following purchase of a new data exchange screen (Seeburger[v]).	
Data exchange transaction via ISDN	Following purchase of a new data exchange screen (Seeburger).	
Dealing with the EURO	Nothing definite.	Yes
Millenium-bug solved	Yes	Yes
Data integration	-only via interfaces -only by the month -high degree of checking (partly by hand) -multiple data storing	Yes Current data positions
Quality safeguard		Yes
Project system		Yes
Changes in the manufacturing process (e.g. Production cells[vi])	Difficult conversion of production due to control of type of parts (e.g. Production cells)	Yes
Modification possibilities of the software by Sidler e.g. stock keeping via laser pistol.	Problematic R&H is not co-operative	Yes Must be programmed to facilitate switching to new releases.
Documentation	Disastrous	Yes
Multi-lingual	Supplementary charge	Yes
Export settlement for Customs, Interstat	Insufficient Must be bought in	Yes
Server platform	AS/400	AS/400, Windows NT, Unix
Transfer to subsidiaries	Yes They require AS/400s if our system is not used	Yes
Consolidation of data (group invoicing)	Explosion of interfaces	Yes
Future partners	No statement as to future development. Capital base? Turnover per head	Financially sound Good future prospects
Future of administrative programs	Continued maintenance for MAS90 is secured up to the year 2001 by selling the servicing to BIW. What follows? By then, IBM should have "San Francisco" on the market and there will be conversion aids.	

Kaltenbach identified the problems with the company's existing EDP software (Figure 7), combined them with the Sidler's needs to meet the new millennium, and questioned to what extent potential vendors could meet the company's needs. Sidler was transferring 500,000 records per month with up to 80 connections per day. Sidler's major customers were also adding growing demands for rapid EDP communication and data transfer, as shown in Figure 8. The investigation focused most heavily on overcoming the weaknesses of the current system. As shown in Figure 9, Kaltenbach listed the core problems

Figure 9 (Continued). Aid to decision-making for future data-processing systems

2. Comparison of investments

		R&H MAS 90, CS	SAP/R3
Seeburger		3,500	
Update of the current quality system	*	100,000	
Purchase of a project system	*	50,000	
Integration of the Sidler modifications	*	50,000	
New programming of the interfaces	*	50,000	
Buy-in of export settlement (PC solutions)	*	5,000	
Software upgrade price to R&H when upgrading hardware	*	15,000	
Training PPS	*	150,000	
Software			608,350
Consultants		110,000	637,000
Hardware AS/400	*	150,000	300,000
PCs and 17" screen	*	150,000	150,000
Total Costs		**823,500**	**1,695,350**

** indicates estimated prices*

confronting Sidler, and addressed the ability of the software supplied by R&H and that of SAP to solve those problems. The results suggested that SAP was a better solution than their current software. Knowing that SAP was a complex software package to use, the team members also visited a number of existing users of SAP (Figure 10). The results of that survey suggested the need for additional staff to help with the implementation of the SAP system. According to one company that was adding to its SAP system, "In order to progress in the network (office communication), one additional person (beginner) for network is needed. Two people are required on a full-time basis for conversion."

The SAP Decision

By October 1997, the decision was made to move to SAP for its added functionality (Figure 11). Rapp had to be absent for its introduction, but had a letter read to his staff concerning the introduction (see Figure 12). He explained the basis of the decision:

We knew SAP was a big elephant, but we had to do something with the new factories in France, Switzerland, and England. We decided to leave the U.S. on their current system since they had other problems to solve first.

We have many products being handled by more than one factory within the group. The purchasing is done from headquarters. We develop the pur-

Figure 10. Summary of SAP visits

1. What has been done so far:

1.1 New functionality of SLIGOS

1.2 New offer: Estimated total costs

Software	608,350
Consultants	637,000
Hardware AS/400	300,000
17" screens for existing PCs	150,000
New PCs	?
Total costs	1,695,350

1.3 Visits to SAP users:	Application	Comments
Seyfert Wellpappe	payroll	SAP is the Porsche, Paisy is the Rolls Royce
Rhone-Poulenc, Lucerne	logistics, PPS	Supplier handling missing, Two employees were hired, team members were relieved from normal duties. Project took 15 months before Big Bang introduction 1/1/97.
Heinrich Schmid, Reutlingen	FI, Control	Using AS400. Has 90 independent companies, domestic and foreign (Sweden, France, Thailand, Switzerland). All will be connected via Intranet. Test was successful. Initial problems with IBM. FI is OK. Production data missing for Control.
United Parts	Project system	FI, Control, wages introduced in 4-6 months with ten projects in real-time operation since 1/97. Development and model constuction includes 50 employees. Every employee is a cost unit and reports daily. It is outsourced into a separate company. Projects are sold to production. Currently, PPS is being introduced. Project head is the key user (a Polytechnic graduate) giving 3 half-days per week. System is being customized with external consultants. Business process in production was reorganized. Master data is input for every business process. Every item is assigned to a process and then copied.

2. What remains unresolved?

Controlling
From LAB to production order, existing offer for test installation.

3. Overall concept for integration of subsidiaries

Intranet
Direct link

4. Introduction of Software

Time Schedule:	Project system
	payroll/time management
	FI
	Controlling
	PPS

chase sources and do the quality monitoring of subcontractors for the French and English operations. We add some further value in the product by building the logistics chain.

Our subsidiary then orders from the subcontractor, does some assembly, includes some parts from Switzerland, and sell it via the French factory because

Figure 11. Pro-SAP arguments

Company size	Datex-P
Available for clients	ISDN
International	ATT
Computer platform	GE
AS/400	Supplier-specific forms
PC-network	Supplier assessment
Connections to the locations; Direct line, Intranet	OS-system
Project execution	Loading-facility management
Data exchange audit trails	Empties management
VDA	EDL management
Odette	Export management
Edifact	Customs
Data exchange links	Intrastat

the customer doesn't want any new supplier codes. If he opened a new supplier code for our German factory, it would add a supplier to his statistics and he has been told to minimize the number of suppliers. We solve that by selling though our French factory. But now everyone tells me that they are not making money on the sales, and hope that someone else does. I would like to find out, from time to time, if we are making money within the group on that product, or not. I can find it out now, but it is a puzzle. That is why we decided to eat this big elephant. That is the only system that gave us the possibility of integrating those subsidiaries. All the other systems were single installations for each factory. That would leave us to do the integration. The greater effort and cost of SAP helps us within the group.

In charging his staff with the task of introducing SAP to Sidler, he further described management expectations:

We all know that our business is no longer possible without data processing. A functioning data-processing system is the backbone of our company. Its integration and flexibility in communication is an important competitive factor.

In project QS 9000, we rethought and rearranged our business procedures. Now we have to copy these procedures in a data-processing system, which will become a model of our company. We must therefore immediately correct any discrepancies we notice.

The introduction of this system must take place within the limits of time and expense. To do this it is necessary to use the best brains our company has for this project. It goes without saying that such people do not have limitless time at their disposal and for this reason we must create time. Everyone involved in this project must consider what tasks they can delegate and how management can help them to devote that necessary attention to this project and still ensure that our customers are in no way neglected.

Think about it. Talk to your superior. Ensure that we are working for the future of our enterprise to the best of our ability without endangering the present.

Sidler signed a contract with the SAP-certified value-added reseller at the beginning of November 1997. The target was to have the SAP installation completed by January 1, 1999. Rapp explained the unexpected problems

We selected SLIGOS-Industries because they were large and had an automotive supplier package. SAP was not very strong in automotive suppliers or in the integration of EDI. We need the German, American, and French EDI standards, since each has something different. That was the beginning of a nightmare, of which we are now in the middle.

In 1997, we obtained QS 9000 certification for all our subsidiaries. We had a project with a consulting company and a new quality manager to prepare our QS 9000 certification in the beginning of 1997. This project was a vehicle for business process reengineering. As we described each business process, we discussed whether it was the right process or whether it could be done better. That was a big project that gave everyone the chance to contribute to the question, "How do I want to make what I do today better?" That was described in QS 9000 with flowcharts. There was a family of 12 parts that represented our processes, beginning with one made here, one from the Czech Republic, etc. We handed this process information to the SAP people to describe how we work and what we wanted. We wanted them to set up the appropriate SAP system. We gave them our "A"- team from the company. I didn't want to be dependent on consultants, so I established our own A team do the work and be able, afterwards, to make the modifications. We wanted their support to connect the first subsidiary into the system, but we would then try to do the second with minimum support.

Figure 12. Dr. Rapp's charge to Sidler staff for the SAP introduction

To Sidler GmbH & Co. **November 24, 1997**

It is over two years now since we launched a project, which, with considerable involvement of energy and expense, saw the integration of the data-processing system then in use. Everyone actively involved in that process knows just how difficult it is to realize integration in practice and what disastrous consequences even the slightest mistake in data input and data maintenance can have. In the meantime we have become used to this situation and are now ready for the next step, which is intended not only to solve some of the problems here in our business, but to take integration a step further beyond the factory gates in Tubingen. We are preparing for the integration of our data-processing systems in the Sidler Group.

Two years ago the first step was necessary in order to break through the limitations set by our previous data-processing system. We could not get any further with our teleprocessing nor did we have the possibility of combining the PPS and controlling systems with each other. At that time, we had already considered introducing a fully integrated interface-free program package, but considering the state of processing know-how as it was then, none of us had the courage -- and it would have been irresponsible anyway -- to throw all the systems overboard and install a totally new PPS system with the existing manpower. So we dug deep in our pockets, demanded additional commitment from you, our staff, to achieve this first stage of conversion from which we were to advance and to learn. And we have advanced and we have learned.

But at the same time we still have, as was to be expected, our problems with the many interfaces, which connect our present systems. We want to change this. At the same time, too, we are no longer producing an ever-increasing percentage of parts ourselves, but share the work with other subsidiaries in our Group without being able always to say exactly whether an article in the Group will be profitable overall or not. This has to be changed.

We have therefore decided once more to dig really deep into our pockets, appealing yet again to your personal commitment and we plan now to introduce a system, which will facilitate this next step. Following the recommendation of an in-house team of experts we have come down in favor of the SAP system and today's meeting sees this introduction.

This particular system is gigantic, but it does everything we need. Having said that, I must warn you not to be tempted to start wanting to use such functions as we would perhaps quite like to have, even if we do not really need them. The SAP-system is capable of overtaxing by far our newly gained data-processing experience and we do not want to risk that. Expert guidance during the introductory phase is intended to prevent this happening, but I would nevertheless ask everyone to be very careful.

We realize that this project will once again put a considerable burden on our key staff. It is like playing a pinball machine: those who try particularly hard and long achieve a high score and win a free turn. That means they can repeat the whole process to win another free turn. And this is what we see as the positive element of this situation: only success will produce such tasks, and only a successful enterprise can set you such tasks, and only by tackling these tasks successfully can we ensure that there will be a basis for the continuation of this success. With this recognition, I do not intend to anticipate the next data-processing conversion, but I am certain that there is life beyond SAP and that there will be more changes in store.

I would very much like to have been with you at the outset of such an important project, but it is my thankless task today to convince a customer that he is doing us an injustice if he has lost faith in our ability to produce quality. I have therefore asked Mr. Krauser to read this statement for me and to oversee the introduction of the new system today. Wishing us all every success with this project.

H. Rapp

Our plan was to go live with our Tübingen headquarters' plant on January 1, 1999, and follow up with our British plant in June 1999. I am afraid that we won't be ready with headquarters by the end of June, to be ready for July 1. That is when the real problem begins, since we cannot fulfill the Y2K problem with the system we have. With all of our automotive customers, we are classified as yellow. That becomes either red or green on July 1. VW came yesterday and will send us the audit organization to check if we are going to become red or green in July. Most of my customers will be coming here within the next several weeks to also determine our status.

The top employees were assigned to the "A" team. They had the personal incentive to learn about leading edge technologies. They are not completely out of day-to-day work, but spent half their time on the SAP project.

Problems with SAP Implementation

In March 1998, German firms discovered that they might have problems with the Euro and Y2K after a German power company failed its Y2K experiment. The SAP partners began experiencing rapid turnover that affected Sidler. Rapp explained:

We started with a good project leader and made good progress until March of 1998. The project leader left the company along with four or five other experts. We then got people who were not experts and started learning again about Sidler. They were trained in only one SAP module. They found that the customizing in the CO module did not correspond to the customization of the MM module and they would not work together. There was no more integrator for the project. They had all left.

We tried to change our SAP vendor, but the market was empty. You couldn't hire anyone. We then tried to hire those experts who had started on our project. There is one still here, but this is the third company he has worked with since he started here. We just pay his new employer for his services. By mid-1998, progress declined. I hired a rare expert for the market price. We now have a project leader to help our consultants get along and tell them how to do their jobs correctly. But we have lost months.

There was no way to complete the project by January 1999. Only one incoming "call off" of dispatch and delivery to one customer has been installed and it worked 90%. Over the last several months, they have been working to develop a concept to move our old database into the new systems. But every time they developed a concept, the systems integration test has failed and has required redefinition of the features and copy programs. It is a nightmare.

They have been customizing SAP since October 1998. We will have to use SAP release 4B, but only release 3 has been available for development.

There was not supposed to be a problem in copying to the new version, but there is a serious problem. Things didn't fit any longer. It was a qualification problem. We tried to get the best supplier, but they turned out to still be weak.

Larger firms have built their own SAP staff and expertise. We tried to do that, but by last year the market dried up after SAP consultants were in high demand. The SAP specialist that I hired has been offered DM 280,000 plus a car. That is no longer realistic for Sidler. I have him in a contract until October, but after than he will likely move on. New consultants have three-days training and are a specialist in only one module. Business processes cross four or five modules.

The SAP system is very powerful. There are so many switches that can be set. There are millions of possibilities for doing things wrong. You do everything at least twice to get it right. We are trying to get the experts from SAP Waldorf to help. If you know them personally, you can get them for a Sunday night, but they are fully booked. It is not a good time to be doing this. We have no choice but to move forward. We will tell everyone that we are working in SAP, even if it is not fully implemented.

Analysis of Sidler GmbH Case Study

This case is about a supplier of internal lighting and trim parts for the automobile industry that is faced with problems in introducing a new ERP system that is capable of handling problems of globalization, introduction of the Euro currency, and Y2K problems of the new millennium. Sidler's managing director, Dr. Helmut Rapp, addressed his problem in early 1997, but has been faced with continuing problems that appear to be beyond his control. The case addresses the decision-making process in selecting SAP as the company's new ERP system, and managing continuing problems with its implementation.

1. **Why is Sidler having so much difficulty getting SAP to run? How would you classify these issues according to cause? Responsibility? Controllability?**

- Consultant turnover has made it impossible to sustain implementation efforts. (This appears to be caused by the demand caused by the Y2K and EURO problems. It is legally the responsibility of the SLIGOS-Industries [now Atos]. It seems not to be within anyone's control.)
- Internal staff capability is too limited to operate without consultants. (Sidler appears to have been conservative in adding staff to manage the SAP implementations. This appears to be management's fault. Today, with the shortage of SAP-qualified personnel, there seems to be little that can be done. Rapp continues to hire one project leader from the fourth consulting firm, and has hired a project expert to oversee the SAP implementation. Hiring more people at this late date may be difficult and counter-productive, unless they are experts.
- SAP appears to be more complicated than expected. Rapp may have relied too heavily on staff to evaluate the needs of SAP. His computer background might have been helpful in determining resource needs and the strategy of implementation. Being a strategic decision, central to management, Rapp might have been more involved. Now that the decision has been made, it appears no longer controllable.
- The July 1 deadline imposed by customers increases the pressure for implementation and stress on the team members. There is little that can be done now, except to offer the team anything they want or need to get the job done. Only Rapp can do this. I am not sure what it might be. More part-time helpers? Shifting of their duties? Discussion with the union to find alternatives?
- The central problem to SAP implementation is that it be addressed on a business process basis. Consultants trained in a module are of little help in assuring that the data streams are correct across the business process. This appears to have been a major problem for Sidler. They have been unable to get any process working, since the data are not correctly aligned across the system.

2. **Was SAP the correct solution for Sidler? Why?**

The evaluation made in making the SAP decision appears to be well done. The fact that only SAP was really evaluated can be questioned. A careful look at the documents will note that Sidler's problems were specifically addressed, in terms of SAP vs. the existing R&H software. With such a narrow comparison, SAP was sure to win. Figures 7, 8, and 9 are central to this discussion.

Figure 8 suggests that SAP is becoming the standard ERP system for Sidler's customers. The applications that customers desire can be accomplished with SAP. What is more important, R&H's software cannot deal with the multiple plants and growing complexity of a global business process. The development of interfaces among operations and functions would be beyond Sidler's ability. In fact, a company like Bosch, who has had its own proprietary operating system, has been forced to move to SAP. The problems faced by Bosch in implementing SAP are similar to those facing Sidler.

The comparison of the weaknesses identified in Figure 7 with Figure 9, the comparison of R&H with SAP in overcoming the weaknesses, makes the decision clear. It might be asked whether a "clean sheet" approach would have revealed a different decision, but with the industry moving to SAP, Sidler is probably have been wise to also use SAP.

3. **What must be done before July 1? What is your strategy?**

While we can review the past decision for ways to improve the making of future decisions, the reality is that Sidler must meet its July 1 deadline for being Y2K compliant for it customers. Sidler has less than a month to make SAP work, after nearly one and a half years of struggles with SAP. What do you do now?

Obviously you work hard to get the system working, but one month is not enough time. What do you do when VW audits your Y2K compliance? Dr. Rapp's strategy was to put on a show of all the windows in SAP that suggest the system is working. Only an expert that wants to test the system would be able to determine the actual situation. You can suggest this strategy, and then discuss the ethics of it. Do you let the company become "red" with your customers, or do you lie?

4. **What contingency plans would you make? What do you recommend to Dr. Rapp?**

You still have a more serious problem to face. How do you handle ordering for the year 2000 requirements? Without SAP working, Sidler has dumped all the forecasts of its customers. Since some of its materials have 26-week lead-times, what recommendations do you make for Sidler after July 1? What actions do you take to get the customer forecasts updated? What do you do if SAP still isn't working by August, the primary vacations time? By September?

There is little choice but to develop some contingency plans. You might determine if there is an accounting firm that can provide help in updating critical records. Sidler cannot accept customer forecasts until SAP is working. The R&H system has already proved to be non-compliant. If you cannot carry out your regular business, and provide expected customer service and JIT delivery, you are out of business. These are not light issues to address.

Chapter Summary

This chapter provided you an opportunity to explore how SAP R/3 was implemented in a company, Sidler GmbH. This case study shows that enormous benefits could be obtained by implementing SAP R/3 in companies, but a lot of planning and expenditures have to be budgeted before the system can be successful. This case study also should convince you that learning more details about ERP systems and becoming an ERP consultant is a good career choice. The use of ERP systems is at its infancy and this market is expected to grow rapidly in the next decade. If you become a certified consultant in one of the ERP packages, you might be able to be a leader in changing many organizations' IT systems. At the same time, as the earlier chapters showed you, you have to learn a lot more concepts and technologies in the future.

Endnotes

i Reprinted with permission by Drs. William R. Boulton and Karl-Heinz Rau, and USW. The Peter Curtius Foundation for USW provided financial support for this case. The numbers in the case have been changed.

ii In contrast to Figure 1, these numbers refer to the 1999 plan.

iii PPS =production/manufacturing planning and control system usually includes demand planning, capacity planning, material requirement planning (MRP), production order processing.

iv CS is a vendor of cost accounting software.

v Seeburger is a vendor for EDI manager software, see: http://www.seeburger.de.

vi Production cells are a local aggregation of different machines and workers to produce a component or a product completely within this center. The logic of the production planning differs because of the autonomy within the center.

Appendix 1:
SLIGOS-Industries SAP Offering

R/3 Automotive Supplier® IBM

The All-Around Solution for Car Component Suppliers

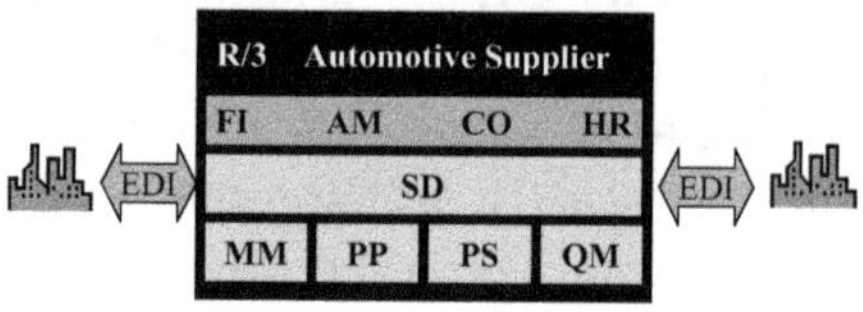

Car manufacturers and suppliers are moving closer together. The reason can be found in the fast changes in the car component supplier market and the aim of car manufacturers to produce and assemble less and less themselves. Logistical contracts are being increasingly reduced and suppliers are already being involved at the development stage of a component part.

The concentration of the systems suppliers on their core competencies and the strengthened link to sub-suppliers requires an integrated approach and optimization of the value-creating processes.

These aims can only be achieved with the help of an integrated software solution tailored to the special requirements of the suppliers.

In partnership with the sector-specialist SLIGOS-Industries, IBM offers a fully integrated solution package, which supports the whole life cycle of a component:

R/3Automotive Supplier® Integrates Business Processes

- from development to purchasing release.
- from inquiry, pilot production to production stage.
- from production launch to end of series.
- from end of series to guaranteed spare parts supply.

Costs and Deadlines

At each stage costs and deadlines can be kept under review and controlled. Costs are taken over into the next processing stage. The total life of a product can be presented at any time in terms of value and amount. The automatic representation of logistic amount flows into value flows of accounting and controlling results in online transparency between budget, calculation, costs, and receipts — important information for decision-making.

Sector-Specific Extensions

R/3 Automotive Supplier® combines the universally applicable R/3 from SAP with the sector-specific extensions for the car-manufacturing supplier industry. Tailored to the needs of the car-manufacturing suppliers, R/3 Automotive Supplier® is the all-around solution for all commercial and logistic tasks.

The sector-specific preset and extension make the introduction of the system easier and support corporate conversion measures. R/3 Automotive Supplier® can be complemented by:

- CATIA for DAD/DAE/CAM.
- EDImanager for electronic data transfer with clients and suppliers.
- ICAM for stock management, control, and production management system.

Appendix 2: Silder Group Companies

European Headquarters

Sidler GmbH & Co.
Bismarckstraße 72
72072 Tübingen, Germany
Telephone (49) 7071/938-0
Fax (49) 7071/938-201

Divisions

SMIG Gmbh & Co. KG
Hauffstrabe 7
72762 Reutlingen
Germany
Telephone (49)7121 38620
Fax (49)7121 38622-2

APAG Elektronik AG
Ringstrabe 14
8600 Dubendorf
Switzerland
Telephone (41)1/8026700
Fax (41)1/8026767

GHS Stanz + Montage GmbH
Siemensstrabe 4
72585 Riederich
Germany

Sidler APAG S.A.R.L.
1, Passage du Clos
92380 Garches
France
Telephone (33)14741 3200
Fax (33)14741 6345

RIVORET S.A.R.L.
B.P. 74
67172 Brumath Cedex
France
Telephone (46)31/412460
Fax (46)31/418422

Germany

Karl Wengeler oHG
Ottostraße 3
50859 Köln
Germany
Telephone (49) 2234/71031
Fax (49) 2234/74422

Sweden

Interopa AB
Kapplandsgatan 96
41478 Göteborg
Sweden
Telephone (46) 31/412460
Fax (46) 31/418422

Great Britain

A.T. Redding
The New House
Malvern Road,
Malvern WR13 5DZ
Great Britain
Telephone (44) 1886/833653
Fax (44) 1886/833654

D.A. Atkin LTD
Cambridge House
3 Newbold Street
Leamington Spa.
Warwickshire CV32 4 HN
Great Britain
Telephone (44) 1926/427591
Fax (44) 1926/315537

Wall & Knight
The Cottage
Chester Road
Castle Bromwich
Birmingham B36 9DF
Great Britain
Telephone (44) 121/7492727
Fax (44) 121/7767067

North American Headquarters

Sidler Inc.
1155 E. Whitcomb
Madison Heights, Michigan
USA 48071-1414
Telephone 1- (248) 583-2710
Fax 1- (248) 583-2723

Divisions

Laotto Manufacturing
11654 East 400 South
P.O. Box 378
Laotto, Indiana, USA
It has added two component molding presses and presses for In Mold Decorating. This QS 9000 facility uses "Impact" programs for continuous improvement.

Chapter XII

Conclusions

Learning Outcomes

- Summarize the SAP R/3 implementation experience of Robert Bosch GmbH and Robert Bosch US
- Review the inventory of skills that you improved by analyzing the case study
- Identify skills that IT personnel will need to possess in future business environments
- List lessons in implementing ERP systems that could be applicable to your future career
- Acquire an in-depth understanding of ERP implementation by analyzing the implementation of SAP R/3 systems at Robert Bosch RB GmbH during 1992-2004
- Analyze the change management processes adopted by Robert Bosch using the change management life cycle theories for the periods 1992-1999 and 2000-2004

Benefits of Using This Book

Robert Bosch GmbH is one of the many companies that are moving toward global forms of organization that transcend national boundaries. However, it was not able to make this move until it had reorganized its information systems, changed the manner in which its factories were being run, standardized its business processes, and chose a divisional form of operation so that the same systems could be used worldwide. Robert Bosch GmbH expects to spend approximately a billion dollars in order to implement the SAP R/3 systems throughout Europe and the world between 2002 through 2008. There are many valuable lessons that can be learned by analyzing the Robert Bosch case study, which we will discuss in this chapter. In addition, we believe you will have enhanced your inventory of skills as you engaged in individual preparation, small group discussions, and large class discussions while analyzing the case study. We will also review these skills in this chapter.

Summary

Chapter I set the stage for analyzing the Robert Bosch case study. Chapter II provided an overview of the structure of a multinational corporation, using the example of Robert Bosch GmbH's corporate structure. Chapters III, IV, and V provided the theoretical foundations of ERP, change management, and database management systems.

We met the heads of the IT division at QI, Robert Bosch GmbH's headquarters, and the heads of the IT division at Robert Bosch US during 1999. Based on these interviews, we developed Chapters VI and VII. Chapter VI discusses the numerous information systems used by Robert Bosch US in order to manage its operations, and examines the pressure to standardize to a corporate-wide information system. Chapter VII highlights the steps taken by QI (Robert Bosch GmbH's corporate IT division), led by Dr. Eggensperger and assisted by Mrs. Bauer, to standardize on the use of SAP R/3 throughout the company. By July 1999, in a period of just over four years, the number of R/3 users in the Bosch group had risen from almost none to 20,000. Don Chauncey, the CIO of Robert Bosch US, came up with a proposal that included a range of architecture options in order to standardize on the use of SAP R/3 through North America. He outlined the SAP direction to the North American Operating Committee during August 1999. They stated that the direction was acceptable, but wanted more specifics, such as costs of "must do items," "a corporate standard," and "need to improve processes." We analyzed the materials provided in Chapters VI and VII and came up with possible responses to the questions raised in those chapters. This information is included as Chapter VIII. This will allow you to compare your analysis of Chapters VI and VII with ours. Please note that there is no "right" or "wrong" answer; you are dealing with a real-world problem and your analysis might be different than ours.

During 2000-2001, there were major changes in the management structure of the IT division. Dr. Eggensperger left the company and Mrs. Bauer moved over to become CIO of Bosch-Siemens-Household appliances; Mr. Chauncey moved to become the vice president of Corporate IT security. Mr. Friedrich took over as the CIO of QI. Mr. Bieganski, CIO of the K1 unit in the U.S., took over as CIO of Robert Bosch US.

We had an opportunity to visit with these new managers during 2004 and obtained up-to-date information as to the progress of the SAP R/3 initiatives at Robert Bosch GmbH and Robert Bosch US. Chapter IX provides a summary of these interviews. The company is planning to spend an enormous amount of money— close to one billion dollars — during 2002 to 2008 to standardize on the use of R/3 systems throughout the company, and expect that they will have about 70,000 users of R/3 system by 2008. The company employs about 1,600 IT employees in Europe, 200 in the U.S., and another 200 in other parts of the world. Of these, about 1,500 employees work on designing the IT architecture and ensuring that it stays aligned to the business processes. Only about 500 employees work on operations and programming aspects of IT. The top management agreed to embark on this project and stay with it, even though the return on investment was not immediately obvious. This chapter also outlines the major challenges facing the IT division as of 2004. Chapter X discusses predictions of how IT might evolve in RB based on Netweaver technologies. Chapter XI presents an additional case study of SAP R/3 implementations in Sidler GmbH.

We believe this longitudinal study of the IT implementation at Robert Bosch and the summary of the implementations at another company provide a wealth of material that will help you in your future career, and will also be useful for managers and top executives in companies who are embarking on implementing ERP systems in their own companies. In the next section, we describe the inventory of skills that you will have exercised as you worked through the material provided in this book.

Possible Improvement in Inventory of Skills

The use of the case study method of learning will lead to the improvement of your skills in many areas. The first three skills — analytical, application, and decision skills — are business-specific and we will thus refer to them as the business skills. The last three skills — communication, social, and time skills — are not business-specific but are valuable to have in general. Hence, we will refer to them as people skills. Creativity does not fit either of these categories and does not seem to be a core skill in either a business or social environment. Nevertheless, it is an important skill to have to separate you from other job applicants; this skill is more like the icing on the cake.

This reasoning leads us to state that the development of students' business skills should be the focus of a business class. In addition, it would be nice to further develop people skills if possible, but each individual student is likely to have plenty of opportunities outside the classroom in a university environment to enhance his or her people skills.

Business Skills

The Robert Bosch (RB) case study is a very good way of enhancing ones' business skills, especially for business students. The Robert Bosch case study requires students to use their analytical skills to understand the underlying business processes. A student must carefully read through the vast amounts of information and focus on the most influential components to get a good understanding of the scenarios the managers faced. Consequently, students must rely on their analytical skills to comprehend the decisions and suggestions made for implementing the SAP R/3 system at Robert Bosch. Then they must apply their decision-making skills to assess which option would be the best implementation for SAP R/3 considering the alternatives and environmental impacts. The third business skill, application, is the missing part needed to fully comprehend the RB case study. This requires the student to have some additional knowledge about ERP systems in general and SAP in particular. Students are encouraged to consult outside resources and apply the knowledge gained to get the most out of this case. Such a real-life case study requires the students to use their business knowledge to understand the decision made.

1. *Analytical Skills*: By analyzing the case study, you have learned to identify the problems facing Robert Bosch's IT division, analyze the four IT structures that were identified, choose one of the IT architectures for implementation, and then defend your selection. This will have improved your critical thinking skills about how multinational corporations operate, their business processes, and how the ERP systems evolved along with the business during the period from 1995 to 2004.
2. *Decision-Making Skills*: The Robert Bosch case study asks you to assess what Robert Bosch US can do and to choose an architecture and discuss how to implement it. By engaging in this process, you learned to discuss the

alternatives, select appropriate decision criteria, evaluate alternatives, choose the best one, and formulate an implementation plan.

3. *Application Skills*: You may have learned about SAP implementation from other books (Kale, 2000), but the Bosch case study made it possible for you to apply the theories learned in those books to a real-world situation.

People Skills

Communication skills, time management skills, and social skills are also enhanced while working through the RB case study. These people skills are also easily acquired outside the classroom if students put forth the effort, as college life offers innumerable opportunities for students to engage in social activities and develop people skills. Students need to realize that nowadays there is a trend towards team work in the business world. Most companies realize that they can achieve greater productivity and faster problems solving by utilizing teams. In general, a real case scenario enables students to acquire a form of experience, the closest form of experience that can be provided outside the actual work setting. Students get a chance to pull all their knowledge together and apply it in a protected school setting, where nobody loses his or her job over a wrong decision. Rather, the RB case provides a safe learning environment.

4. *Communication Skills*: The chapters provided you with details of the statements made by IT managers at Robert Bosch GmbH, and the slides and proposals they developed. You had the chance to analyze these communications and to critique them while finding answers to the assignment questions. If you had the opportunity to write your own report and present it to a class, then your communication skills would have further improved.
5. *Time management Skills*: By reading the cases, you can see the pressure on the IT managers to deliver solutions within strict timelines. In addition, you must read through, understand, analyze, and form recommendations about the problems posed by the case study in a relatively short time — a matter of weeks. This forces you to schedule your educational activities carefully, especially if you had to coordinate meetings outside your regular class time with others in your group, and manage time effectively.
6. *Social Skills*: If you were able to participate in small group and class discussion in analyzing the case study, you learned about how to work effectively as part of a team and to deal with your peers. This learning includes conflict resolution skills and practicing the art of compromise.

Because so much of your future work will involve committees, task forces, boards, or project teams, learning to work effectively in a group will benefit you significantly.

Creativity

Incorporating a role-play situation into the curriculum for the Robert Bosch case study allows students to express their creativity and work with one another in a different manner. Students have had the opportunity to get to know the key figures making the decisions at the company, learning about Robert Bosch at a different level than that presented in the case study:

7. *Creative Skills*: You had to be creative in understanding and analyzing the Robert Bosch case study. At the same time, you will have become aware that the process involved in implementing an ERP system in another company will have many similarities and many differences. We hope you have exercised your creative skills in analyzing the Robert Bosch case study and will use it further in your career.

Suggestions for IT Students

The skills expected of IT students have changed drastically in recent years and are frequently not reflected in the courses that you take as students. Therefore, as we worked through this case study, we asked the IT managers at Robert Bosch GmbH and at Robert Bosch US to make recommendations on what skills they consider would make you a valuable employee in the future. They suggested that in addition to acquiring strong technical skills (necessary, but not sufficient), you need to acquire eight different skills. In addition to listing their suggestions below, they are illustrated with instances from the case study where such skills were used by an RB engineer or manager.

1. **Think logically and be analytical.**

You need to be able to reflect on and identify the problems facing the organization, use structured methods to analyze these problems, focus on the most critical areas, and use logic and analysis to choose between alternatives.

In the Robert Bosch case, a major problem they faced was how to implement the ERP system. At one extreme, the implementation could be done by introducing all the applications/systems in every unit simultaneously. The other extreme was a step-by-step implementation. Each of these could either be implemented as function-oriented or process-oriented approach. The management constructed an architecture consisting of each of the four possible outcomes. By analyzing the architectures, the company was able to determine the optimal strategy by assessing where the sum of both categories was minimized.

2. **Think in terms of business processes and how IT could improve them.**

IT has to be designed so as to improve business processes and make the lives of users better. The information technologies, reports, and high-tech gadgets are simply the tools that will be used to meet these needs, and this should be taken into account when designing systems.

The management of Robert Bosch was able to build a system to improve business processes and make the lives of users better. Robert Bosch made the move to a company that transcends national boundaries by reorganizing its information systems and standardizing its business process.

3. **Improve your communication skills so you can communicate technical concepts effectively.**

It is critical to understand what the end-users are telling you and to be able to explain technical issues to them, enabling you to structure the IT system so that it meets the real needs of those users. Without effective communication, many systems cannot be successfully implemented.

The members of the Bosch organization were in constant contact with one another during the implementation of SAP R/3. Robert Bosch is a global company with a global strategy. It is vital to be able to communicate with other members of your organization. The foreword to this book notes that a survey of 539 IT heads from a range of industries revealed that the most important problem facing the CIOs was to "find the best practices for partnering with business units and delivering the greatest value to the organization." This is a prime example of the need for effective communication in order to understand the business processes related to the company's operations. Don Chauncey was under pressure from the HR and IS staffs, who were asking to update their systems to PeopleSoft to cover their payroll and human resource needs. However, Krish Kumar was evaluating the information systems of the company to address its customers' needs. They knew they had to come together and establish a unified strategy to cope with their growing and changing company. They faced many issues while making the transition to a standardized global information system, and effective communication skills allowed them to accomplish their goals of implementation of SAP R/3.

4. **Learn to understand the impact of what you study and connect theories learned with practice.**

It is important for you to think about what you are learning and how it will have an impact on your future. It is critical to learn new academic theories, but they have to be matched with the needs of practice. This is important for industry.

5. **Understand different cultures, since you may be working on multicultural teams.**

Students who understand different cultures, can speak multiple languages, and have traveled abroad are very valuable in a multinational corporation.

The SAP R/3 global implementation was being used in the areas of accounting, logistics, and human resources. Bosch is successful with this global approach because of the QI department; it retains a dedicated internal team of consultants that deal specifically with the implementation and use. They ensure that the master data is standardized and that basic structures and templates are used. Gunter Lehman was responsible for ensuring the synergy of the structure and harmonizing the SAP implementation. Taking into consideration that national companies are acting for multiple divisions that have country specific regulations and business environments, he created the RB/3 system architecture that established the standard guidelines that would be followed in each country.

6. **Be tolerant and open to ideas from others.**

Multinational corporations employ people from many very different cultures and it is important for them to be able to work together and exchange ideas. It is also important to match people to the tasks they are especially suited for. Bieganski states, "I will pick on one of my staff members from Germany. We tell him you are going to plan until you die, at some point in time you have to put this plan into action. Until it is perfect, he may not move. In contrast, we may act without a large planning process. Therefore, we let him lead the engineering part of the project since planning is very essential there."

The case shows that this lesson is difficult to practice. Jurgen Schlaefer defined his personal motto as "unrest produces creativity." Mrs. Bauer stated in the QI Newsletter of July 1999 that more than one person had declared that they were not ready for change. Comments included, "Our plant is already optimized" and "...just another thing that makes life more complicated." She believed that standardization should involve getting together with the experts to determine and introduce the best process for a certain procedure or the best system for a certain product.

7. **Be a very good team player, since ERP implementations are complex and involve people from all disciplines.**

Many of the ERP projects at Robert Bosch are led by people from the end-user organizations. It is critical to be a good team player and provide leadership to the team when needed.

Robert Bosch had to make the evolution from a multidomestic company towards adopting a global strategy. In order for this evolution to be successful, all members working together can accomplish the goal more easily than each person working alone.

8. **Learn how to design IT architectures, a critical skill for the future.**

About two-thirds of the IT employees at Robert Bosch work on designing the IT architecture, a core competency for the company. Programming is a key competence (20 % in-house and 80% outsourced to other countries for Robert Bosch GmbH), but other IT skills are considered to be standard competences that can be easily outsourced. Therefore, you must ensure that you have sufficient educational background on designing IT architectures and then obtain as much practical experience as possible by working on internships or projects.

The architecture design team was a valuable asset to Robert Bosch. The success of the project can be attributed to a well-designed system that provided the company with the standardization and flexibility it desired.

In the next section, we summarize the lessons that can be learned by studying the Robert Bosch case study, both as observed by us and as stated by the IT managers at Bosch.

Lessons Learned by Analyzing RB Case Study

There were several lessons learned by the IT managers at Robert Bosch and the authors during this longitudinal study that would be of use to IT managers, top management, and IT students. These are:

1. **Be committed to the ERP implementation project and be a leader.**

While top management will set the direction, successful implementation requires a serious commitment from all involved. An ERP project is not just an IT project, or a finance project, or a logistics project. ERP is a whole organizational project that involves many people, and in the Robert Bosch case, over 70,000 employees. An effort of this magnitude mandates that many will become corporate, divisional, and plant-level leaders. Leaders will guide organizational change, design business processes, identify possible failure points, and make difficult decisions. Through their leadership, an ERP project can be successful.

ERP implementation projects tend to be multiyear endeavors and consume a huge amount of a company's resources. For Robert Bosch GmbH, the SAP R/3 implementation began during 1995 and is expected to be complete by the end of 2008. The expenditure from 2002 to 2008 alone will be over a billion dollars as the SAP R/3 system is rolled out worldwide. The QI team has developed a detailed roll-out plan and checks the status of the implementation against the plan periodically.

It is difficult to provide a return on investment (ROI) computation for the SAP R/3 implementation since it is hard to compute the return from the individual divisions. It is also difficult to compute what is saved through the integration or harmonization of data. It is much easier to compute how many people were employed or how much was paid to consultants. As the SAP R/3 projects are being implemented, Bosch is beginning to be able to collect some information about the ROI, although it is still difficult to compute the figures before the implementation begins in each area. The implementation is being performed to achieve three major objectives, but these are not necessarily quantifiable. Harmonizing data structures is a very important gain. If Robert Bosch wants a complete picture of all sales to BMW, previously the company had to sort through hundreds of records to generate the figures; with the harmonization of data that is achieved by SAP R/3, it is now a much easier and faster process and the answer can be delivered the same day. The next benefit is harmonization of the processes. Once a problem has been solved for a process in one plant, then the same problem in another plant can be solved very quickly. This advantage accrues from standardization of processes across the company, leading to reduced complexity in the system since there are fewer systems and fewer interfaces. The structure is still complex, but it is now very lean. Therefore, the overall SAP R/3 system has become a lot less complex to manage compared to the multiple information systems previously utilized within the company.

In order to implement the SAP R/3, it has been necessary for the IT division to behave like a benevolent dictator. At the same time, it is critical to get the support of top management for such major IT initiatives.

2. **Cultivate top management support.**

Implementing an ERP system is not a grass roots effort. Rather, ERP systems require strong leadership from the top. The top includes the Board of Directors, CEO, CFO, and CIO. Their strong leadership determines much of the successful implementation. They must be united in their resolve, financial backing, and patience. With a financial commitment of $1 billion over six years, top management at Robert Bosch will become very demanding. Leaders should all understand the future benefits and be committed to business change. With such commitment, an implementation can be stable and consistent, and weather leadership change.

Major discussions were held with the chief financial officer and important members of the board at Robert Bosch GmbH throughout the implementation process. The success story of SAP R/3 implementation in an automotive business provided a good starting point in convincing the board about the need for the one billion dollar investment during 2002-2008. Only 20-50% of the cost is directly for the IT people, hardware, or software, depending on the size of the site. Most of the investment goes into training the end-users and standardizing the processes. It also helps that the CIO of Robert Bosch GmbH sits on two boards, the IT board as CIO, and the board that supervises definition, implementation, and roll-out. This makes it possible for a uniform opinion of QI's strategies to be communicated to the board, thereby ensuring that there is no miscommunication. Change management is a major issue and it is important to work with the business divisions in implementing change.

Top management made a decision during 1995 that no more programming would be done in-house to build specific systems; from that point on, SAP R/3 would be the standardized software used by all divisions within the company, effectively moving from 80% home-grown software and 20% purchased software to 100% purchased software. This decision forced many of the IT people to face the reality of the direction being taken by the company and change their aspirations.

Once the top management understands the benefit of the ERP system in terms of how it harmonizes the processes within the company, they become more involved in the SAP R/3 implementation and begin to talk in SAP R/3 language. Consider the issue of the level at which a company code should be provided, which became an important question that required top management approval. A company code is the lowest legal entity for external reporting purposes on the balance sheet, P/L statement, and so on. Therefore, the assignment of company codes defines the reporting level in the organization. For example, if several different product lines (belonging to multiple divisions of the company) within a plant have individual company codes, then the reports would be produced by product lines and at divisional level; there may be no reports produced at the plant level. This is a major change for the plant manager; his or her role is redefined to be equivalent to that of a "hotel manager" who ensures that the plant runs smoothly, security is appropriate, and employees are comfortable in their work environment, instead of being a "plant leader" who ensures that all the product lines are producing according to targets. The division is now responsible for a product, from the purchase of parts, through manufacturing and warehousing, to sales. A plant is thus now divided into many different product lines. Even though

IT allows for such a change to be implemented in a company, the initiative to do so has to come from the top management.

The top management is also forced to rethink its assignments to middle-level managers. Earlier, conflicts used to occur because top management would ask the plant manager, as well as the division manager, to maximize the production of certain parts. This led to conflicts between the plant and division managers as to who was in charge of producing the product in question. With the company code being assigned to the division, the responsibility now rests squarely with the divisional manager for ensuring that the production is up to norms and for coordinating work with the plant manager to ensure that the workers are provided with sufficient resources to produce the needed products. The use of the SAP R/3 system therefore makes explicit inconsistencies in job assignments and helps the organization change effectively.

The top management has found that understanding the IT architecture has been helpful as the company diversifies operations or purchases other companies. The divisional approach does not constrain the management according to which country the plants are located in. This makes it possible for Robert Bosch GmbH to fulfill customers' needs quickly and at the most reasonable cost. They have also begun to use SAP R/3 terms in their meetings, indicating a major buy-in from the top management.

3. **Focus on life cycle management of IT projects and products in the company; plan for new ones as the old ones become obsolete.**

A company has to think about the life cycle of a product and plan what product will replace it after a particular amount of time. This is similar to buying a car: You may buy a car intending to keep it for five to six years and then replace it after that time period. Consequently, you will be aware of the investments to be made and plan for capital investments at an appropriate time. In the same way, you invest the amount of time and funds needed for a particular IT project; you design it with obsolescence in mind. It is important to have a road map of where the systems are headed.

4. **Designing IT architecture is a core competence and business design has become critical; this needs to be the basis of the project plan.**

Robert Bosch's implementation of a single ERP system throughout its organization stemmed from its need to have its IT systems aligned with its business processes, and also to control costs. It is important to have a clear system architecture so that the different IT projects fit together to make a coherent whole. Such architecture makes the relationship of a particular project to the overall architecture clear and transparent. Many new harmonization tables had to be developed at Robert Bosch GmbH. For example, a round steel rod is bought based on its weight in kilograms, but used in centimeters length. Therefore, the system needs to include a conversion table from kilograms to centimeters for that material. If it is to be used in the U.S., a ratio of kilograms to inches is also needed. Thus, harmonization tables have to be developed worldwide and also have to be implemented. Any mistakes in the table will have a worldwide impact, but once the error is found it can be fixed at one step. In the earlier systems, this type of information was not harmonized, leading to errors in reporting and logistics functions. It is critical that IT personnel are able to understand these technical issues, develop harmonization tables, and communicate them to the users worldwide.

Because businesses are not stable and keep changing, designing a business is a difficult process. This is not like engineering or product design, where the parameters change slowly. There are two ways in which IT architecture can be designed: (a) multiple systems, where the client states: "the business is changing and our IT has to follow suit. We will provide whatever resources are needed to make the IT follow the business. We are not worried about expense." In such cases, an ERP system may not be necessary, as they can afford to develop and implement multiple duplicative systems; (b) single ERP systems, which are much more effective for an organization that wants to control IT costs such as Robert Bosch and expects the IT to be aligned with the business processes. This approach is more effective if the company wants more flexibility with lower costs. ERP systems could be used to manage change in the company and could become an enabler of change management.

It might be also important to differentiate between two kinds of IT developers: those who develop platform software and those who develop application software. Those who want to work on platform development are likely to find jobs in ERP companies, such as SAP, Oracle, and so forth, while those who want to do application development will be those who will succeed at Robert Bosch GmbH. It is important for you to decide where you fit in as an IT professional and pursue the appropriate career path.

5. **Emphasize disciplined execution; focus on a few projects and roll them out fast.**

It is not possible for you to do everything. You must focus on a few projects and ensure that they are implemented well. Otherwise, you might be working extremely hard, but not achieve the goals needed. This implies that once target dates are agreed upon for particular IT project implementations, they have to be kept. It is important to plan the project, manage expectations, and communicate so people know what to expect.

6. **The ability to communicate technical specifics to a general audience is critical.**

IT leaders in an ERP project communicate with many important audiences. First, they must communicate effectively with top management. Top management often listens to ERP vendors with eager ears as they tout financial pay-offs. However, IT leaders should also help translate these promises into realistic implementation requirements. The ability to explain complex technical specifics to a multidisciplinary top management team will empower them to make better decisions. IT leaders should also discover the reporting requirements of middle managers. Understanding their business and production needs is critical for providing actionable reports for decision-making. IT leaders need to communicate with the general employees as well. The design team and trainers will be the ones making an ERP system "real" to the employees. Their effective, sensitive communication will ease the transition for all users.

Being a CIO is a challenging assignment; it requires that person to work with internal processes and people, rather than working with customers who are outside the company. Frequently, it is easier to work with customers than with your coworkers inside the company when you need to be a champion of change management. It is important that the CIO be able to communicate with the board, particularly with regard to technical concepts. At the same time, it is essential for IT managers to learn about the organization as a whole, its functions, divisions, products, and the manufacturing operations. Otherwise, they may not be able to work with divisional level managers in harmonizing the business processes. Implementing an SAP R/3 system implies changing the management process and at the same time implementing a massive amount of software in a plant or a division.

7. **Control of technology management. ERP systems are large enterprise systems.**

They can be formed and shaped during the design phase, but once they have been installed, changes in the business process or architecture are very difficult. As a result, the design and implementation of IT architecture is critical. Some companies make this a core competency, while others contract this out as a one-time service. Regardless of the path, inefficient business design or poor architecture will impact the entire corporation. Duplicating existing, dysfunctional business processes will perpetuate inefficiencies and doom a successful ERP implementation.

The expenses involved in implementing SAP R/3 are huge and generally require new computers, hardware, and software. It is important to manage these technologies so as to achieve the project's objectives. QI developed a customized version of SAP R/3 for the divisions. It took more than half a year to convince the plants that there was no need to customize the software for each plant and that the customized versions had to be used in a standard manner within the divisions. The company found that they had to develop a system and no longer rely on in-house software. The IT system for Robert Bosch GmbH is effectively an organism that supports the various processes used in the company using prepackaged, standardized software.

When IT people build systems in-house, they become specialized in particular hardware/software and are then rarely able to transition to other platforms. The company found that the use of SAP R/3 made many of its employees much more valuable to the organization, since they could be deployed anywhere in the world to support the company's operations. It is critical for IT people to know basic programming software, but they may not need to use those skills on a day-to-day basis.

8. **Employees need to be able to work effectively in multidisciplinary, multinational, and multicultural teams.**

Robert Bosch GmbH is a multinational company employing people from many different countries. It is essential for the employees to work effectively alongside each other, while at the same time being aware of cultural and country differences. A company such as Robert Bosch, being a German company, values people and does not hire and fire them casually. This is in sharp contrast to some of the American companies that they merged with, whose value system is centered around producing at a cheaper cost and who do not mind hiring and firing personnel. In such cases, it becomes critical for the IT personnel to understand the cultures of the people in the acquired companies and work well with them in order to accomplish the company's objectives.

Implementing an information system is not a painless process for most corporations and with it comes numerous trials and tribulations as these corporations seek ways to utilize their investment efficiently. It is very important to realize that the Robert Bosch case study merely shows the strengths and weaknesses of the ERP implementation for one firm, and these may or may not be applicable to all firms. Even though many companies may buy the same software, the final working software package will vary considerably between firms because the corporate culture and set-up of the implementing company will dictate the way the product is actually installed. Another point to note is that the business climate that the company operates in will also dictate the degree of success that is possible in the initial implementation process. The cultural background that the business operates in, and the managerial paradigms that are associated with this, speak volumes about the employees' work ethic and style. These can factor greatly in the success, or lack thereof, when implementing ERP systems.

In the next section, we discuss the change management life cycle discussed in an earlier chapter and apply that framework to analyze the RBUS case study.

Relating the RBUS Case Study to the Change Management Life Cycle

We analyzed the case study presented in chapters VI and VII (1992-1999) and chapter IX (2000-2004) in order to understand how RBUS coped with the changes that took place. The findings from this analysis are presented in this section.

Changes During 1992-1999 (Chapters VI and VII)

Phase 1: Diagnosis: Before 1992, RB and RBUS used multiple information systems to manage their businesses throughout the world. During 1994, the Board of Management came up with a major agenda for change by defining two mission statements for the IT management team: The IT division has to operate like an external IT provider by meeting or exceeding the benchmarks of the providers, and it had to be a leader in setting policies, procedures, and standards for IT throughout the organization. The Board of Management probably created this mission statement in response to market and industry changes. RB's managers realized that they needed more timely and accurate data in order to make better decisions for the organization and to remain competitive in a global market. This recognition of new, more efficient, competitors can be related to the threat of new entrants in Porter's five forces model. During the same time period, the automotive industry pushed for a tighter integration of suppliers to create a more efficient value chain. This driving force can be associated with the bargaining power of customers in Porter's five forces model. RB had to improve the integration of its corporation in order to stay competitive.

In addition to the industrial forces acting on the company, there were several market forces, such as globalization of markets, flattening of organizational structures, growing influence of information management, and the growth of e-commerce, which might have influenced the decision of the Board of Management. This complex interaction of outside forces required QI to strategically align RB's IT infrastructure with the strategic business future RB managers foresaw.

Once the Board of Management had set the business requirements, it was up to QI to implement and support the company's future through IT. These management observations would fall under the diagnosis stage of the Change Management Cycle.

Mrs. Bauer remembered that in 1994, "[QI] started looking for a new ERP system." QI seems to have been convinced that an ERP system would have the capability to support the future business needs of RB effectively. Since 1994, it has been clear to the IT managers and staff at Robert Bosch that a change in technology was required in order to stay competitive and fulfill the requirements of the Board of Management. Between 1994 and 1998 additional problems and requests surfaced, resulting in an official decision to change.

Phase 2: Change Decision: The gap between RBUS's current situation and its future needs became large enough that it was clear to the company's IT managers that a final change decision was necessary. In 1998, Don Chauncey, Krish Kumar, Mike Bieganski, and Mrs. Bauer were the ones responsible for making that change decision. The ultimate decision had been forced upon these managers, partially as a result of decisions made by the board of management and its suppliers, Computron, which had announced that it planned to discontinue its technical support of the company's existing systems.

Krish explained how the different systems made it difficult to integrate the operations and meet customer demand effectively:

To manage the customer demand precisely, we have a very sound physical layer, we have a very sound network layer, but we do not have a very sound application layer. That's what we need to focus on in the future to meet our objectives.

Because of the current nonintegrated system solution, we have created pockets of implementation gaps between the functional departments, sales, engineering, manufacturing, distribution, logistics, and financials. Not only do we have complexity in the organization, but also the old reporting methods add time to the production of performance information. So management requires us to position the application layer to address business growth and increased complexity.

Further, our customers are asking us to provide both total systems and modules. They want to reduce the number of suppliers. This brings more complexity. We have to work closely with the OEMs. The whole system or module is integrated into the automobile, and we have to bring in our tier-2 and tier-3-suppliers and even our competitors at times. We need to ensure that the design integrates with the other components of the car.

Don talked about the issues that were requiring changes to be made to the payroll and human resource system:

Though we had changed every other system in the company in five years or less, we had not changed the payroll/human resource system. In fact, our payroll system is an old mainframe-based system. It has a human resource component, but the users indicate that the system doesn't have the human resource capabilities that are required. The vice president of corporate HR wants very urgently to implement a new human resource system. About a year ago he requested a study for a new system.

After discussions with Computron, they finally agreed that they would support the AS/400 version of their software until the end of 2001. They have other customers who had Y2K issues and they want to support their customers through Y2K so as not to further add to the workload.

The end of this phase includes Don Chauncey's presentation to NAOC and their reluctance to give the start sign for the project without further clarifications.

Phase 3 – Change Planning: The fact that it is difficult to exactly define the transition into phase 3 is due to RB cycling back and forth through these stages of the change management life cycle.

Planning the SAP implementation clearly falls under phase 3, while the final decision-making by NAOC is part of phase 2. Nevertheless, by July 1999, QI had come up with a detailed implementation plan. The change plan followed an incremental top-down approach to implementing the ERP system. The managers would ultimately decide how the process would look that was to be implemented, and their implementation would be mandatory unless it would result in extensive losses to the division.

Phase 4 – Creating Readiness: The initial processes were developed in small groups and did not include any input from the majority of employees. During July 1999, the RB/3W initiative was launched to get the company-wide implementation of SAP/R3 started. At this point RB had made several good choices regarding the change management life cycle. RB thoroughly analyzed the problem, realized that technology can be used in addition to changing RB's business processes to become more efficient, and had the full backing of the management staff.

Phases 5, 6, and 7 – Change implementation, change reinforcement, and change evaluation: The only unit in RBUS that was able to implement SAP R/3 was the K1 division under the leadership of Mike Bieganski. Seven modules were installed at 11 sites and a total of 1,200 users were impacted. This implementation was considered a success. An important point is that Mike did not agree with the strategy recommended by Don and questioned the implementation approach for RBUS. He wanted to build SAP in one big instance in the U.S., so everybody would use the same configuration based upon the QI SAP model. His evaluation of the change process was that, "We don't have an RBUS harmonization model with RB in Germany." In addition, Don reported, "K1 wanted a lower cost allocation for HR." It is clear from these discussions that the CIOs of the different divisions had not been able to come up with a common plan for designing and implementing SAP R/3 at RBUS.

Changes During 2000-2004 (Chapter IX)

Phase 1 – Diagnosis: It is interesting to note that the management of IT operations at QI and RBUS had changed completely by the end of the implementation at RBUS. This was possibly due to the pressure by top management to embark on the change process to implement SAP R/3 throughout the organization and a perceived lack of effective implementation by 2000. The new CIO, Mr. Friedrich, became a CIO after being a board member at a subsidiary company of RB. Mike Bieganski, who had implemented SAP R/3 at K1, was made the CIO of RBUS. Harald Fontius, who had implemented SAP R/3 in RB's Brazilian plants and had worked with Mike in implementing SAP R/3 during 1992-1999, became the vice president for program management.

Phase 2 – Change Decision: It was clear during this time period that the top management had made a commitment to embark on the use of SAP R/3 throughout the corporation and was willing to invest close to one billion dollars without a formal return on investment computation. Mr. Friderich explains, "During 2001, Bosch was in a position that they were at the end of the life cycle of their major IT systems. They had to invest in IT systems in any case so as not to hurt the business, and that was the reason that the board of management accepted this investment without a sound ROI calculation."

Phase 3 – Change Planning: Mike Bieganski reflects the new change atmosphere in RB as he states that "The major success factor in accomplishing a project of this magnitude was to focus on the business process model instead of the technology." In addition, cross-functional teams became an important tool for designing the new processes. These teams ensured that people and processes were aligned. Gerd Friedrich, the new CIO, focused on three change management efforts. First, he changed the management focus from a production focus to a value chain approach. This resulted in products and processes that were aligned. Second, he increased the importance of process harmonization, resulting in the alignment of people and processes. Last, he emphasized that only one development system was going to be rolled out. Even though the divisions would get their separate R/3 instances, this would ensure that RB could work with the same harmonized codes throughout the organization. Finally, management also decided on one ERP system architecture, using an "Enabling Layer."

Phase 4 – Creating Readiness: By 2004, the whole company was aware that the new company-wide ERP system would have an impact on the employees. The new management attitude that put people first in the change management efforts indicated that employees received the attention they needed to make the ERP implementation a success. Gerd Friedrich illustrates the new corporate culture, "[the] company wide discussion about values, targets, and respect for their employees. In this context, there exists a common understanding that everyone will follow decisions that are made corporate-wide and will implement the decisions. Having this common understanding makes it easier to implement the changes and systems." This is supported by his statement that "less than 20% of the total costs for the ERP systems are spent by the IT implementation organization; the funds are used by the other two organizations to create change in the company, change business processes, educate employees, and so on." The IT management had realized that majority of the investment has to be devoted to creating change and sustaining it.

Phase 5 – Change Implementation: The implementation of IT systems at RB in Europe covered 70% of the business, with 30% covered by mostly harmonized systems on a regional level. The systems in Mexico and the U.S. used the same implementation approach as in Europe. For example, there was only one separate R/3 instance for all of North America for HR-payroll. The business processes in the consumer goods sector were consolidated and integrated into the SAP R/3 instance. The industrial sector had its own SAP R/3 initiative. RBUS was working with QI so that they could generate regional views, but at the same time create a global model that reflected global processes.

Phase 6 – Change Reinforcement: The board had adopted policies that enforced the change to the SAP R/3 system. Mr. Friedrich states, "The different divisions were invited to send representatives to work with the first group in order to define the business processes. If they chose not to participate, they were forced to accept the processes as they were defined." He further states, "The board does not accept excuses not to use the standardized and harmonized processes. If a division is not willing to accept the standardized processes, we will bring it up to the board and then the division has to show figures that make it clear that their business will be damaged if they use the standardized processes. It is not only a question of whether they like the standardized systems or not; they really have to show the negative impact on the business by using a standardized process. The board has come up with such a process to resolve disagreements, but I have to admit that up to now no division has used this escalation process."

Phase 7 – Change Evaluation: The change has had a major impact on the company by changing the focus from plant to divisions, increasing the importance of process harmonization, and developing one system for the rollout. The fact that the IT team got funding to the tune of one billion dollars to complete the implementation shows that the top management has confidence that the standardized system will be implemented across the company.

The change in the system requires that everybody involved shares the same understanding of terms and processes throughout the company. Therefore, employees must be trained to understand the importance of using the same terminology throughout the organization. In a multiple instance scenario, there are technical limitations to interfacing with each of the databases. Under multiple instances it will also take more time to convert the different datasets resulting from their different data structures in order to combine them to create a common body of knowledge. Both approaches will have advantages and disadvantages, but we believe that a single instance would benefit RB in the long run. The major challenge in implementing a single instance is to align the company structure and processes to become one organization in the mind of the employees. Such a change in the corporate culture will allow for a successful implementation of a single instance ERP system. One should also consider that a single instance can be less labor intensive to manage by preventing the need for RB to maintain multiple datacenters. Having a single instance can provide cost advantages and management advantages to RB if well implemented.

Unfortunately, a single instance also creates a single point of failure and Robert Bosch needs to carefully implement a contingency plan to address this vulnerability. It would be a serious problem for RB if its ERP system were not operable 24/7. RB thus needs to think about physical as well as virtual data security and back-up plans. Such a contingency plan should include a hot data center that can be switched online at any time. A physically secure building should be used to protect the IT equipment from physical break-ins, as well as natural forces such as water, heat, and so on. In order to ensure 24/7 operations for RB, the datacenter should have its own generator and conditioned power lines. Virtual security includes proxy servers, firewalls, routers, and encryption to disguise and protect the stored data from intruders. The database will reflect the structure of RB and the job responsibilities held by the employees. By using a well thought-out combination of indices and views, RB will be able to restrict the amount of information viewable by the end-user while still storing everything in one central location. It is important that the security issue be considered seriously as the single SAP R/3 system becomes widely used in RB.

We hope to revisit RB in 2010 to evaluate the final impact of this long and complicated change process. It will be interesting to hear the company management's view of the whole process and to see how it has affected the company's bottom-line.

Chapter Summary

We have learned a tremendous amount by performing a longitudinal study, spread over ten years, on how SAP R/3 is being implemented at Robert Bosch GmbH. This book provides you, the reader, with an opportunity to experience the implementation experience using a case study format. We encouraged you to work individually, in small teams, and in large groups to analyze the case study and gain an appreciation of the experience of Robert Bosch GmbH with its SAP R/3 implementation. We want to conclude with a parting thought provided by IT managers at Robert Bosch GmbH and Robert Bosch US, "You can learn the basics about IT design and architecture in school; but you have to work in companies to internalize the ideas and to become competent IT designers." The business processes and IT systems used by multinational companies are large and complex billion-dollar systems. Therefore, IT divisions want their employees to be able to design and implement their IT architectures successfully. This implies that in addition to basic technical competence, you have to develop business design skills, a knowledge of IT architecture, good communication skills, team working skills, leadership skills, and the ability to garner top management support so that you can succeed in the IT profession. We wish you the best as you acquire these skills by continuous learning throughout your career, and success in your life.

References

Kale, V. (2000). *Implementing SAP R/3: The guide for business and technology managers*. Indianapolis, IN: Sams Publishing.

About the Authors

Chetan S. Sankar is the Thomas Walter professor of management at Auburn University, Alabama, USA. His research on improving engineering and business education by using research methodologies has been well recognized by the award of six National Science Foundation grants totaling $1.5 million. He brings firsthand knowledge of the real world from outreach activities to enhance research and education experiences of students. He directs the Laboratory for Innovative Technology and Engineering Education for Business Students (LITEE-B), and his work has had significant impact on improving the higher-level cognitive skills of more than 10,000 engineering and 7,000 business students. He has published more than 70 articles in referred journals, including the *MIS Quarterly, Management Science, European Journal of Engineering Education, Decision Sciences Journal of Innovative Education, Journal of STEM Education: Innovations and Research, IEEE Transactions on Education, Information Technology and Management, Journal of AMCIS, Journal of Engineering Education, Journal Creatividad Education Y Desarrollo, Project Management Journal, Decision Support Systems, The Academy of Management Executive, Engineering Management Journal,* and *the Journal of Global Information Technology Management.* He was the third-place winner of the Society for Information Management 1990 Paper Award, first-place winner of the 1995 Decision Sciences Instructional Innovation Education Awards Program, the 1997 American Society for Engineering Education Instructional Unit Award, and 1999 American Society of Mechanical Engineering Curriculum Innovation Award. He has extensive experience in developing and publishing award-winning case studies based on issues faced in real-world businesses.

Karl-Heinz Rau is a professor at the Business School, Pforzheim University, Germany. He teaches management information systems, object-oriented system analysis, application systems in management accounting, and information management. He is a visiting professor at University of Ljubljana, Slovenia. He was the head of the Department of Business Information Systems within the Business School during 1994 to 1999. He was a professor of business information systems at the University of Cooperative Education, Stuttgart, during 1988 to 1992. He was a consultant to IBM Germany during 1984 to 1988. He obtained his doctoral degree from the Tuebingen University during 1980 to 1984. He was a member of planning staff at IBM Germany from 1976 to 1980. His university is one of the first ones to be associated with SAP University Alliance Program in Germany starting in 1992. Dr. Rau has published a book on business planning and another on office automation. In addition, he has published more than twenty research articles in journals and books in the area of information systems and management accounting. He is a coeditor of the *Gabler Wirtschaftsinformatik-Lexikon* (Encyclopaedia of Business Information Systems), and a coeditor of the book series *Praxis der Wirtschaftsinformatik.* Dr. Rau is a member of the Gesellschaft für Informatik e.V. (Association of Computer Science). He has taught SAP R/3 applications and systems for more than ten years and has accumulated expertise in the SAP R/3 environment by working with companies.

* * * *

William Boulton is the C.G. Mills professor of strategic management in the College of Business, Auburn University. His research is focused on global, technology-based competition. Dr. Boulton was a visiting professor at Germany's premier executive development institute, Universitatsseminar der Wirtschaft at Schloss Gracht, in Erftstadt during 1999. He was a visiting scholar at the prestigious Institute for Fiscal and Financial Policy of Japan's Ministry of Finance in 1993. He has also been a visiting professor at Keio Business School in Japan in 1992, 1993, and 2001, where he was also hosted as a Fulbright Research Scholar in 1986. Since 1992, he has conducted a number of research studies for U.S. government agencies on electronic manufacturing technologies and industrial development policies of Asian countries. The studies he edited include Electronic Manufacturing and Packaging in Japan (1995), Semiconductor and Electronic Manufacturing in the Pacific Rim (1997), and Information Technologies in the Development Strategies of Asia (1999). Dr. Boulton also chaired and edited the Electronic Manufacturing in Hong Kong and China study for the International Technology Research Institute in 2000. Dr. Boulton is the author of numerous articles and cases covering topics of competitive strategy, technology and innovation management, boards of directors, and strategic planning. He also developed *The Resource Guide for the Management of Innovation and Technology* for the AACSB in 1993, and wrote a case and textbook, *Business Policy: The Art of Strategic Management* in 1984.

Gerd Friedrich holds a degree in mathematics and physics and spent the first six years of his career in the research and development department at Siemens AG. He then moved to Nixdorf, a fast-growing computer company in Germany, after which he joined the telecommunications division of Bosch, starting in R&D and ending as a board member before he was appointed as the head of the QI management team and as the CIO of Robert Bosch GmbH in 2000.

Helmut Krcmar holds the chair for information systems at the Department of Informatics, Technische Universität München (TUM), Germany, since 2002. He worked as postdoctoral fellow at the IBM Los Angeles Scientific Center, as assistant professor of information systems at the Leonard Stern School of Business, NYU, and at Baruch College, City University of New York. From 1987 to 2002 he was chair for information systems, Hohenheim University, Stuttgart. From 2000 to 2002 he served as dean of the faculty of business, economics, and social sciences. His research interests include information and knowledge management, IT-enabled value webs, service management, computer-supported cooperative work, and information systems in health care and e-government.

Valentin Nicolescu has worked at the University Competence Center at Technische Universität München, Germany, since 2003 as technical director and scientific assistant. Besides the technical administration of a large computer center, he is concerned with integrating the new aspects of SAP NetWeaver in teaching. From 1998 to 2003 he studied economic science at Hohenheim University, Stuttgart.

Daniel C. Pantaleo, Fulbright Scholar, Dean, Academic Vice President, and Provost, spent four years as vice president for product development with an interactive multimedia education and training developer after a career of 25 years in higher education as a faculty member. Dan left that post in 1997 to assume the responsibilities of program manager for SAP America's University Alliance Program, USA, where he has been able to apply his knowledge of higher education and interactive learning to significantly develop the program. Expanding his responsibilities, he was appointed director of higher learning initiatives for SAP America, with responsibility for several higher learning projects including the University Alliance Program throughout the Americas. Dan was appointed to lead the SAP Innovation Institute as its vice president in October 2000. The SAP Innovation Institute was responsible for identifying and facilitating collaborative research projects with universities, an Executive Education Program, the University Alliance Program, and for identifying innovation strategies, methods, and best practices. In July 2002, Dan became part of the SAP Global Communications organization with global responsibility for the University Alliances Program. Recognizing the opportunity to expand the development of strategic content and research for SAP, the University Alliances research and innovation team was joined with the strategic analysis team of SAP Global Communications to form the SAP Strategic Issues Group. Dan heads the project development and management of this new team. Dan holds a BS degree in chemistry from Manhattan College and a PhD in inorganic chemistry from Emory University. He served as a Fulbright Scholar in science education in the Philippines, held a postdoctoral appointment in the chemistry department at Louisiana State University, and took postdoctoral coursework in astrophysics at Georgia State University. Dan is a published author and researcher and a member of several professional associations, including the Innovation Council of the Conference Board, and serves on several advisory boards and councils.

Holger Wittges has been the executive director of the SAP University Competence Center at Technische Universität München, Germany, since 2004, and works there on his postdoctoral lecture qualification. Before, he was project manager at a large German telephone company. He attained his doctorate in economic science at Hohenheim University, Stuttgart, on workflow and business process design.

Index

C

D

E

L

M

N

O

P

Q

R

S

T

U

V

W

X

Y

www.ingramcontent.com/pod-product-compliance
Lightning Source LLC
LaVergne TN
LVHW062113200926
848484LV00007B/1655
9781591407768